AF566677

# Checking Experiments in Sequential Machines

# Checking Experiments in Sequential Machines

**ASOK BHATTACHARYYA**
Assistant Professor
Electronics and Communication Engineering
Electrical Engineering Department
Delhi College of Engineering
Delhi, India

JOHN WILEY & SONS
New York Chichester Brisbane Toronto Singapore

First published in 1989 by
WILEY EASTERN LIMITED
4835/24 Ansari Road, Daryaganj
New Delhi 110 002, India

**Distributors:**

*Australia and New Zealand*:
JACARANDA-WILEY LTD.
GPO Box 859, Brisbane, Queensland 4001, Australia

*Canada*:
JOHN WILEY & SONS CANADA LIMITED
22 Worcester Road, Rexdale, Ontario, Canada

*Europe and Africa*:
JOHN WILEY & SONS LIMITED
Baffins Lane, Chichester, West Sussex, England

*South East Asia*:
JOHN WILEY & SONS, INC.
05–04 Block B, Union Industrial Building
37 Jalan Pemimpin, Singapore 2057

*Africa and South Asia*:
WILEY EASTERN LIMITED
4835/24 Ansari Road, Daryaganj
New Delhi 110 002, India

*North and South America and rest of the world*:
JOHN WILEY & SONS, INC.
605 Third Avenue, New York, N.Y. 10158, USA



**Library of Congress Cataloging-in-Publication Data**

Bhattacharyya, Asok.
Checking experiments in sequential machines.

Bibliography: p.
Includes index.

1. Fault tolerant computing. 2. Sequential machine/switching theory. 3. Electronic digital computers—Testing
I. Title.
QA76.9.F38B47 1989 004.2 88–33863

ISBN 0-470-21365-5 John Wiley & Sons, Inc.
ISBN 81-224-0113-9 Wiley Eastern Limited

Printed in India at Urvashi Press, Meerut.

**To**

MY MOTHER

*Srimati Haripriya Devi*

AND MY FATHER

*Late Dr. Brajendra Nath Bhattacharyya*

# Foreword

With ever increasing complexity of digital system design and in view of enormous time and cost involved in realising the design, fault detection is becoming as important as the design effort. In fact, fault detection needs to be integrated in the design philosophy itself. Unfortunately, the fault detection has not received due attention and by and large there is a degree of lack of awareness of its basic features. A part of it may be ascribed to the scattered literature on the subject.

It is in this context that the present monograph on the fault detection of sequential machines will be considered as timely.

The essential features of the monograph are: comprehensive review of the existing literature, system classification of the field into well sequential chapters and illustration of applications. The author who has his own contribution in the field has provided an organisation of the subject in a manner such that the material can be used as a reference source in advanced courses.

It is my belief that the monograph will find ready acceptance by the researchers in the field of digital system design.

PROF. A. B. BHATTACHARYA
Centre for Applied Research in Electronics
Indian Institute of Technology Delhi
New Delhi 110 016

# Preface

In the present era of very large scale integration technology, the topic of Fault-Tolerant Computing, an important discipline of computer technology, is recognised almost everywhere. Fault-tolerant computing can be defined as "the ability to execute specified algorithms correctly regardless of hardware failures or software errors", whereas the technology of fault-tolerant computing encompasses theory and techniques of fault and error detection and correction, modeling, analysis, synthesis and architecture of fault-tolerant systems and their evaluation. Under the broad spectrum of the problems of fault tolerant computing, the design of self-diagnosable computers or computers that are easily testable and repairable comes immediately to the mind. All these require an excellent knowledge of fault testing and isolation techniques as they pertain to the devices with which modern computers are built.

The process of applying tests in determining whether the computer is fault free or not is generally known as fault detection or checkout. The present book imparts a systematic study of the problems of fault detection and designing of checking experiments in sequential machines. With the ever increasing demands of integrated circuits and the application of modules in modern digital system, it has become extremely important to evaluate from terminal experiments whether a sequential circuit operates correctly. This had started developing with the works of E. F. Moore as early as in 1956 and is continuing till this date, as its importance is being felt while testing even random access memories or bit-sliced microprocessors. Of late, considerable attention has been also focussed on the problem of designing and implementing logic circuits that are easily testable. Simultaneously, more and more efforts are also being made to design sequential machines with built-in efficient fault detection capabilities of easily testable and diagnosable nature.

Assuming that the readers must have the basic background of switching theory and logic design, the present book incorporates a methodical development of the problem of checking experiments in sequential machines and takes into consideration the development of various easily testable sequential machines with built-in efficient fault detection capabilities. Also, the applications of checking experiments in developing the test sequences of random access memories and of bit-sliced microprocessors are being considered. Apart from this, the book has been provided with an appendix which deals with salient features for testing microprocessors/ microprocessor-based systems.

This monograph is compiled from the various works of the eminent persons which are available in scattered form in different journals/books in isolated manner. The purpose of the book is to make available at one source the widely-scattered information pertaining to the fault detection problems in sequential machines. It is hoped that the book would benefit those who are interested in the area of fault-tolerant computing, and in designing and testing of the integrated circuits and modules.

The author takes this opportunity to express his indebtedness to his dear Professors Late Dr. A. K. Choudhury, of Computer Science/Radiophysics and Electronics Department of the University of Calcutta, India and Dr. S. R. Das of Department of Electrical Engineering, University of Ottawa, Ontario, Canada, for their encouragement.

Also, the author feels great pleasure in thanking Dr. A. B. Bhattacharyya, Professor, Indian Institute of Technology, New Delhi, for going through the manuscript of this book and for his valuable forwarding comment regarding this monograph.

The author also feels pleasure to mention that it was the firm insistence and encouragement of his wife that enabled him to complete the work.

ASOK BHATTACHARYYA

# Contents

# Chapter 1

# Introduction

In recent years, the design of digital systems with fault detection capabilities has assumed significant importance because of increased emphasis on fault-tolerant system design. The technology of fault-tolerant system design encompasses both theory and techniques of fault detection and fault correction, modeling, analysis, synthesis, and architecture of such systems as well as their pertinent evaluations. Digital systems broadly fall in two distinct categories: combinational systems and sequential systems. It is well known that in combinational or acyclic digital systems, the system output at any instant is solely dependent on the system inputs at that instant, whereas in sequential systems, the system output at any instant depends not only on the present value of the system inputs but also on the past history.

Two different approaches have been taken into consideration to generate tests for fault detection of a combinational logic circuit. The first approach is basically an algebraic approach and it employs the Boolean equations of a circuit as a basis and conducts a structural analysis on its function. In most cases special expressions are proposed to replace the Boolean equations and based on these expressions, test generations are possible. The works by Poage [169], Seller et al. [196], Boosen and Hong [28] and Clegg [47], are among those using this approach. The second approach for test generation in a combinational circuit is the path sensitizing method introduced by Armstrong [10] and improved later by Schneider [195] and Roth et al. [185] among many others. The idea behind this technique is common in test generation though different approaches have been used in applying it to a circuit. For detecting and analysing a fault on a particular line, a test must be found that establishes a path from this line to one of the primary outputs. The procedure is called a path sensitization technique. The *D*-algorithm introduced by Roth [184] is a procedure to generate a test, employing path sensitization technique. To find a test for a given fault on a line, the *D*-algorithm first establishes a sensitized path through this line and then it desensitizes to obtain a consistent set of primary inputs to support this sensitized path. This algorithm guarantees the generation of tests for detectable faults in any combinational circuit. Improvements and modifications have been made by Roth et al. [185], Bouricius *et al.* [30], and Su et al. [213] and others.

For a combinational circuit, the test generation is a complex task and many constraints have been placed on the circuit to simplify the problem. Most of the circuits have only a single primary output. Reconvergent fan-outs are found to be a problem and, in some studies, are not allowed at all. Generally, redundancy has been assumed to be nonexistent. Detailed results have been obtained for circuits with limited levels and inputs. These results are valuable because an understanding of simple parts can lead to solution of the whole problem. Because of the current technology of large-scale integrated circuits, logic circuits tend to have a large number of gates. As a result, the Boolean equations and special expressions used in the first approach become extremely complex and unmanageable. Although significant insight and elegant results have been gained, this approach is becoming impractical. In the second approach, logic diagrams of networks also become complex as the number of gates increases. The second approach is still more practical because it assigns a value to a line according to the gates immediately connected to it; therefore, complexity does not affect the decision making elements. In addition, computer-aided logic design is a well-established practice. Design-automation systems are gradually replacing the designer's task of generating tests for their designs. Logic diagrams rather than Boolean equations serve as inputs to such a system.

The fault detection in the case of sequential circuit is also developed following two distinct approaches. In the first technique, called the circuit testing approach, an experimenter requires exact knowledge of the details of the circuit realization, and also of the faults that might possibly occur. The circuit can be modeled by a pseudo-combinational circuit with a special structure in such an approach. By this transformation, most of the techniques used for generating test sets for combinational circuits can be readily extended to handle sequential circuits. This kind of transformation is applied properly in synchronous sequential circuit. In the case of asynchronous sequential circuit because of the critical timing factors involved, such transformation finds hardly any application. The circuit testing approach is thus recommended for relatively small circuits with thoroughly known circuit diagrams. This approach provides optimal experiments for detecting and locating permanent single or multiple stuck-at-zero (*s-a-0*) and struck-at-one (*s-a-1*) faults in any sequential circuit.

The approach most generally followed for fault detection in sequential circuits is the transition checking approach, which does not require any knowledge of the actual circuit realization, but does assume the knowledge of the desired transitions. This approach detects, in addition to the class of stuck-at faults, the class of failures that cannot occur from only *s-a-0* or *s-a-1* faults in the circuit, but are of unknown origin. Though in sequential machines, the problem of fault detection and diagnosis is rendered more complicated due to the presence of memory elements yet this latter approach based on terminal observations cover a larger class of malfunctions. E.F. Moore [157] and F.C. Hennie [113] did pioneering work in this area. In developing his approach of fault detection, Hennie [113] has defined certain types of input/output sequences, and showed how, subject to the limitations in the structure and the faulty behaviour of the sequential machine, these sequences could be used to assist in the derivation of a checking experiment, to verify if (a) all the machine states exist, and (b) all defined state transitions can be made. In this kind of transition verification approach, the experiment is designed to take the machine through all its state transitions in a way that a definite conclusion can be reached as to whether or not the machine is operating correctly. In general, such a technique is complex and rather difficult, being dependent on many factors, like (i) the total number of states of the machine, (ii) whether the machine is reduced and strongly connected, (iii) whether the machine possesses distinguishing sequences, and (iv) whether a fault that has occurred belongs to a class of known faults etc. Hennie also showed how distinguishing sequences can be used to considerably reduce the total length of a checking experiment. Unfortunately, all reduced state tables do not possess simple preset distinguishing sequences. Although this fact does not prohibit the derivation of a checking experiment, it surely renders the problem more complicated. To overcome this difficulty, recently Kohavi and Lavallee [133], Murakami et al. [160] and others have demonstrated how a state table that does not possess distinguishing sequences can be effectively modified by adding extra input or extra output logic to achieve an easy testing of the machine. These approaches have opened up new avenues in solving the problem of checking sequence design in sequential machines.

Of late, considerable attention has been focussed on the problem of designing and implementing logic circuits that are 'easily testable'. Reddy [183] has defined an easily testable machine as one having the following attributes: (i) small test set, (ii) no logical redundancy, (iii) test set is derivable without much extra work, either during the design phase, or after the circuit has been defined, (iv) structure of the test set is such that it is easy to generate and also to interpret the results, (v) faults can be located to the desired degree. This list is general only; the other properties may also be necessary, and thus can be added to the list, viz. (vi) limited number of final gate-count and this number is not excessively high as compared to a 'normal' implementation, (vii) minimum number of additional primary control inputs, and observable outputs required for enhancing testability. In recent years following the same line of approach, designing of the checking experiments in sequential machines for various easily testable machines are being considered. Development of various easily testable sequential machines with built-in efficient fault detection capabilities are now being dealt simultaneously and checking experiments are being designed for them.

This book is a monograph only. Though a methodical development of checking experiments in sequential machines is being presented yet it has not been possible to include all research works that have been done in this field. Some important contributions towards development of checking experiments in sequential machines are discussed in detail; while the principles of a few other methods are briefly given. Other works are referred.

This book is organised as follows. In Chapter II, certain important terminologies for the various problems of state identification experiments in sequential machines are discussed. The characteristics of certain special types of sequential machines developed for better testing purposes are also considered. In Chapter III, a comprehensive review of the work done on the problem of fault detection in sequential machine is presented. Here, the attention has been given to the work of Moore [157], Hennie [113], Kime [128], Gönenc [96], Hsieh [120], Sheng and Das [206] and others. Chapter IV reviews the work done for developing checking experiments in sequential machines that use the concept of machine modification either with extra inputs or extra outputs. Here, the motivation of the work of Kohavi and Lavallee [133], Murakami et al. [160], Holborow [116], Fujiwara and Kinoshita [84] and that of Fujiwara et al. [85] are discussed. In Chapter V, recent approaches developed for designing checking experiments in sequential machines utilizing the principles of machine modification with the combination of extra inputs and extra outputs are discussed.

Apart from the earlier discussion presented in Chapter II for the various state identification experiments, in Chapter VI the transition matrix approaches for solving the measurement and control problems in sequential machines are discussed [61, 63, 200, 92]. Finally, in Chapter VII two specific applications of the checking experiments in sequential machines are given. Here at first, the recent technique of fault detection for the random access memories by developing checking experiment is discussed as per the scheme of Hayes [104]. Also the application of checking experiments in testing bit-sliced microprocessors is discussed as per the schemes that have been developed recently by Sridhar and Hayes [211–212]. As the fault detection in bit-sliced microprocessors has been presented in Chapter VII, an Appendix which covers briefly other interesting aspects of testing microprocessors or the microprocessor-based systems is included.

# Chapter 2

# Terminology

## 2.1 Introduction

As more and more digital circuits are being manufactured these days in encapsulated or integrated forms, analysis of behaviour of finite state machines from terminal measurements alone is assuming significant importance. The sequential machine to be experimented upon is considered to be available to the experimenter simply as a 'black box', with only accessible input and output terminals. Application of an input sequence to the input terminals of the machine, along with an observation of the corresponding output sequence produced at the output terminals is commonly referred to as an *experiment*. Major objective of such an experiment is to furnish an experimenter with the following information: (i) what state the machine was in at the start of the experiment, (ii) what state the machine comes to at the conclusion of the experiment, (iii) whether the machine is operating correctly, and so on. The state identification experiments for sequential machines are designed to identify either the unknown initial state, or the final state of a machine. A brief discussion on the properties of certain major state identification sequences as used in the design of fault detection experiments in sequential machines will be given in this chapter. Also definitions of certain special types of sequential machines possessing special diagnosing sequence are included.

The experiments in sequential machines can be classified broadly into the following categories: (i) *adaptive experiments*, in which the inputs applied to the machine at any instant of time depend on the information provided by the previous outputs, and (ii) *preset experiments*, in which the entire input sequence to be applied to the machine is predetermined independent of the outcome of the experiment. In any case, the *length* of an experiment is the total number of input symbols applied in conducting the experiment and the *order* of the experiment is said to be the number of input subsequences (i.e., the sequence separated by decision-making operations) with which the experiment is composed of. A preset experiment is an experiment of order 1 and the adaptive experiment is an experiment of order 2 or more. Because, in an adaptive experiment, the applied input sequence is composed of two or more subsequences, where each subsequence, except the first is determined on the basis of the responses resulting from the preceding subsequences.

Generally, one machine is said to be a copy of another machine, if both machines have identical transition tables, and if both are at the same state before the experiment commences. Experiments are performed usually on a single copy of the machine and such experiments are known as *simple experiments*. A more general type of experiment consists in applying input sequences to a number of identical copies of the same machine, each copy being assumed to be in the same initial state at the start of the experiment. Experiments involving several identical copies of the same machine are called *multiple experiments*. The simple as well as the multiple experiment can either be preset or adaptive. The design of multiple experiments seems to be of limited application in sequential circuits, since in practical situation, it is difficult to obtain several identical copies of the same sequential circuit, being in the same unknown initial state. Simple experiments are, in general, preferable to multiple ones, since simple experiments are rather easier to perform though they do not always solve every state identification problem. The *multiplicity* of an experiment is said to be the number of copies of the machine under investigation that are required for conducting multiple experiments. A simple experiment is an experiment of multiplicity 1, and a multiple experiment is an experiment of multiplicity 2 or more.

For the simple state identification problem, the sequential machine $M$ which is with the experimenter can initially be in any one of its $q$ states and the *initial uncertainty* regarding the states of the machine is the set $(S_1, S_2, \ldots, S_q)$. The set of states $(S_1, S_2, \ldots, S_q)$, one element of which is, to the experimenter's knowledge, the initial state of the machine $M$, is often called the admissible set of $M$ and usually denoted by $A(M)$. The states in $A(M)$ are called the admissible states. In general, the *uncertainty* or *ambiguity* regarding the states of $M$ after the application of an input sequence $X$ is a subset of the $X$-successors of the states contained in the initial uncertainty. A collection of uncertainties is known as an uncertainty vector, and the individual uncertainties contained in the vector are called the components of the vector. The trivial uncertainty vector contains components having a single state each while a homogeneous uncertainty vectors have components containing either single states or identical repeated states. For a given initial state uncertainty, a successor tree displays graphically the $I_i$-successor uncertainties for all $I_i$, where $I_i$ is an element of the input alphabet of the machine.

EXAMPLE: Consider the flow table of a machine $M_1$ as shown in Table 2.1*. The successor tree for $M_1$ is shown in Fig. 2.1.

Successor tree for machine $M_1$ with an initial state uncertainty $(S_1S_2S_3S_4)$ contains four levels. Each branch is labelled with the input symbols which

*In all the chapters, the sequential machines are given the names $M_1$, $M_2$, $M_3$, etc. They are to be identified by their respective Table numbers and the concerned chapter.

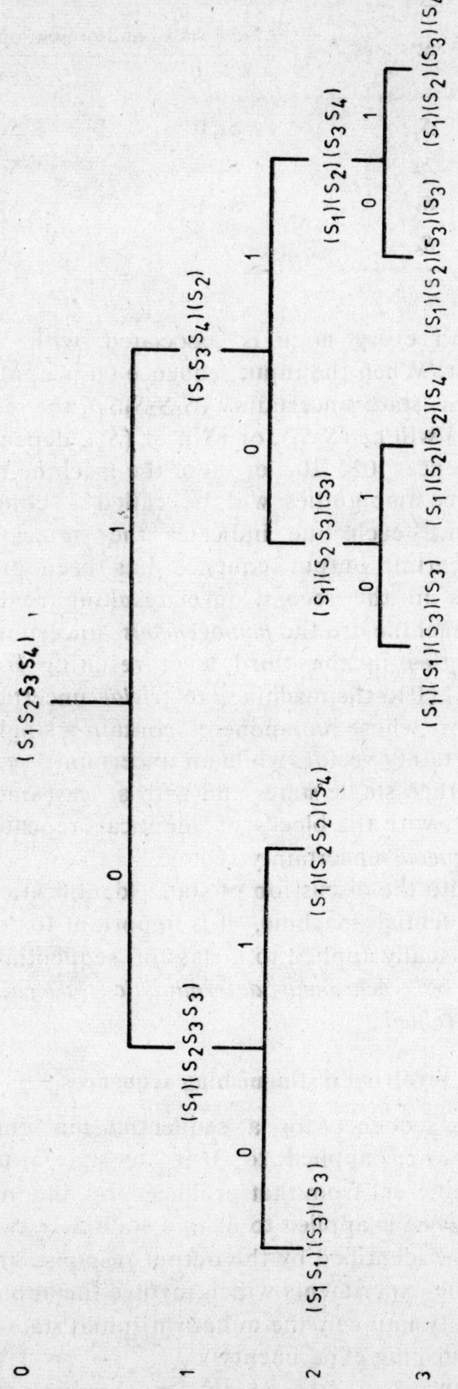


Fig 2.1 Successor tree for machine $M_1$.

Table 2.1 Machine $M_1$

| Present states | Next states and present outputs | |
|---|---|---|
| | $I_1 = 0$ | $I_2 = 1$ |
| $S_1$ | $S_3$, 0 | $S_4$, 1 |
| $S_2$ | $S_3$, 0 | $S_1$, 1 |
| $S_3$ | $S_1$, 1 | $S_2$, 0 |
| $S_4$ | $S_2$, 0 | $S_3$, 1 |

it represents, and every node is associated with the corresponding uncertainty vector. When the input sequence 00 is applied to the machine $M_1$ having initial state uncertainty $(S_1S_2S_3S_4)$, the resulting state uncertainty (ambiguity) will be $(S_1S_1)$, or $(S_3)$, or $(S_3)$, depending on which of the output sequences 01, 10, or 00 of the machine has been produced. Each of these three ambiguities will be called a conditional ambiguity (uncertainty), since each one indicates the state ambiguity under the condition that a certain output sequence has been produced. The first two uncertainties in the second level resulting from the application of 00 or 01 to the machine are the *homogeneous* uncertainty vectors, while all the uncertainties in the third level resulting from the application of 100, 101, 110, 111 to the machine are *trivial* uncertainty vectors. The uncertainty vector whose components contain a single state each is said to be *trivial* uncertainty vector, while an uncertainty vector whose components contain either single states and with at least one block of identical repeated states or, with the blocks of identical repeated states only, is said to be *homogeneous* uncertainty vector.

Before going into the discussion of state identification experiments in detail for a sequential machine, it is important to emphasize that such experiments are usually applied to a class of sequential machines which are assumed to be *synchronous*, *deterministic*, *reduced*, *strongly connected* and *completely specified*.

## 2.2 Experiments involving distinguishing sequences

A distinguishing sequence for a sequential machine $M$ is an input sequence which when applied to $M$ in any state $S_i$ produces an output sequence that is different from that produced by the machine when the same input sequence is applied to $M$ in a state $S_j \neq S_i$. Thus the state of the machine can be identified by the output response to the distinguishing sequence. The experiments which involve the application of an input sequence to identify uniquely the unknown initial state of a machine are said to be distinguishing experiments.

A path in the successor tree of the machine gives a distinguishing sequence, if and only if, it starts in the initial uncertainty and terminates in a node associated with a trivial uncertainty. Thus in Fig. 2.1, there are four distinguishing sequences of length three for the machine $M_1$, viz., 111, 110, 101, and 100.

So far, we considered only the preset experiments in which the choice of each input symbol is predetermined and is not influenced by the response of the machine to the preceding input symbols. But a distinguishing sequence can be adaptive also, in which the choice of each input symbol is determined on the basis of machine's response to the previous inputs. Consider a machine $M_2$ as shown in Table 2.2 with its adaptive distinguishing tree shown in Fig. 2.2. The complete experiment can be devised as follows:

Table 2.2 Machine $M_2$

| Present states | Next states and present outputs | |
|---|---|---|
| | $I_1 = 0$ | $I_2 = 1$ |
| $S_1$ | $S_3$, 0 | $S_1$, 1 |
| $S_2$ | $S_4$, 0 | $S_3$, 1 |
| $S_3$ | $S_2$, 1 | $S_4$, 1 |
| $S_4$ | $S_3$, 1 | $S_1$, 0 |

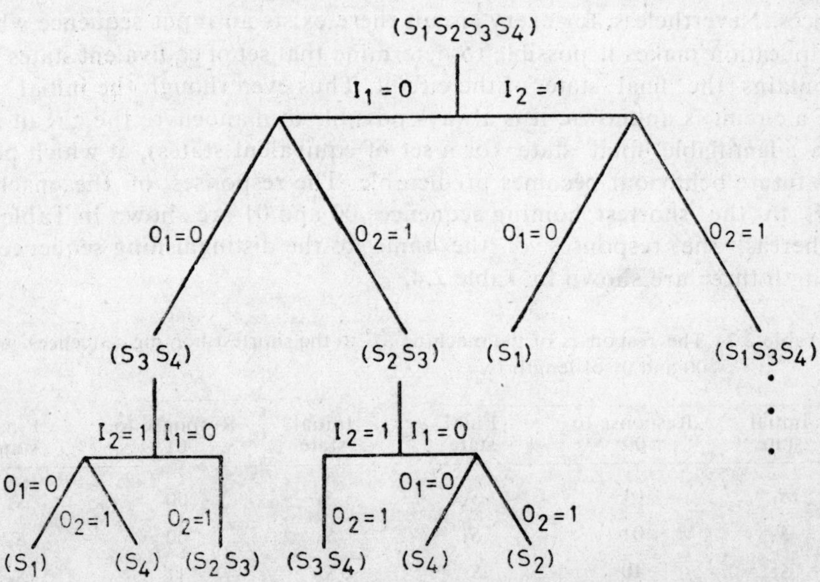


Fig. 2.2 Partial successor tree for the adaptive distinguishing experiments for machine $M_2$.

*Step 1*: Apply an input 0 and inspect the corresponding output. If the output is 0, go to Step 2; if it is 1, go to Step 3.

*Step 2*: Apply an input 1. If the response is 0, the initial state was $S_2$; if it is 1, the initial state was $S_1$.

*Step 3*: Apply an input 0. If the response to this input is 0, the initial state was $S_3$; if it is 1, the initial state was $S_4$.

The advantage of the adaptive experiment is that they are relatively short; in addition, there are machines for which only adaptive experiments can be constructed. In an adaptive experiment, the output is to be inspected after the application of each input, which generally slows down the entire experiment.

## 2.3 Experiments involving homing sequences

A preset homing sequence for a given sequential machine is an input sequence whose application makes it possible to determine the final state of the machine by observing the output sequence the machine produces. A homing sequence can easily be obtained by noting a path in the successor tree leading from the initial state uncertainty vector to a node having a trivial or a homogeneous uncertainty vector. For the successor tree as shown in Fig. 2.1, the shortest homing sequences are 00, and 01. But the sequences 100, 101, 110, and 111 are also homing sequences, even though, none of these sequences is the shortest sequence. As the latter four sequences are also distinguishing sequences, thus every distinguishing sequence is a homing sequence, though the converse is not true. Every reduced sequential machine has a preset homing sequence; circuits containing equivalent states may or may not have homing sequences. Nevertheless, for every circuit there exists an input sequence whose application makes it possible to determine that set of equivalent states that contains the final state of the circuit. Thus even though the initial state of a circuit is unknown, it is always possible to manoeuvre the circuit into an identifiable final state (or a set of equivalent states), at which point its future behaviour becomes predictable. The responses of the machine $M_1$ to the shortest homing sequences 00 and 01 are shown in Table 2.3 whereas, the responses of the same to the distinguishing sequences of length three are shown in Table 2.4.

Table 2.3 The responses of the machine $M_1$ to the shortest homing sequences viz., 00 and 01 of length two

| Initial state | Response to 00 | Final state |
|---|---|---|
| $S_1$ | 01 | $S_1$ |
| $S_2$ | 01 | $S_1$ |
| $S_3$ | 10 | $S_3$ |
| $S_4$ | 00 | $S_3$ |

| Initial state | Response to 01 | Final state |
|---|---|---|
| $S_1$ | 00 | $S_2$ |
| $S_2$ | 00 | $S_2$ |
| $S_3$ | 11 | $S_4$ |
| $S_4$ | 01 | $S_1$ |

Table 2.4 The responses of the machine $M_1$ to the following four distinguishing sequences of length three, viz., 111, 110, 101, and 100

| Initial state | Response to 111 | Final state | Initial state | Response to 110 | Final state |
|---|---|---|---|---|---|
| $S_1$ | 110 | $S_2$ | $S_1$ | 111 | $S_1$ |
| $S_2$ | 111 | $S_3$ | $S_2$ | 110 | $S_2$ |
| $S_3$ | 011 | $S_4$ | $S_3$ | 010 | $S_3$ |
| $S_4$ | 101 | $S_1$ | $S_4$ | 100 | $S_3$ |

| Initial state | Response to 101 | Final state | Initial state | Response to 100 | Final state |
|---|---|---|---|---|---|
| $S_1$ | 101 | $S_1$ | $S_1$ | 100 | $S_3$ |
| $S_2$ | 100 | $S_2$ | $S_2$ | 101 | $S_1$ |
| $S_3$ | 000 | $S_2$ | $S_3$ | 001 | $S_1$ |
| $S_4$ | 111 | $S_4$ | $S_4$ | 110 | $S_3$ |

## 2.4 Experiments involving synchronizing sequences

A synchronizing sequence for a given sequential machine is an input sequence whose application is guaranteed to leave the sequential machine in a certain final state, regardless of the particular initial state of the machine and the output sequence being produced. The application of a synchronizing sequence for bringing the machine into a specified final state constitutes synchronizing experiments. Only some machines possess such synchronizing sequences, while most others do not.

EXAMPLE: To illustrate, let us consider the synchronizing tree of Fig. 2.3, for the machine $M_3$ as shown in Table 2.5.

Table 2.5 Machine $M_3$

| Present states | Next states and present outputs | |
|---|---|---|
| | $I_1 = 0$ | $I_2 = 1$ |
| $S_1$ | $S_2$, 0 | $S_4$, 0 |
| $S_2$ | $S_1$, 0 | $S_2$, 0 |
| $S_3$ | $S_4$, 1 | $S_1$, 0 |
| $S_4$ | $S_4$, 1 | $S_3$, 0 |

In the tree in Fig. 2.3, since we are interested only in the final state regardless of the output, an uncertainty like $(S_1S_2S_3S_4)$ is simplified as $(S_1S_2S_4)$, avoiding the repeated entries of states. The input sequence 01010 which is described by a path in the tree leading from the initial state uncertainty to a singleton uncertainty ($S_4$) i.e., uncertainty containing just a

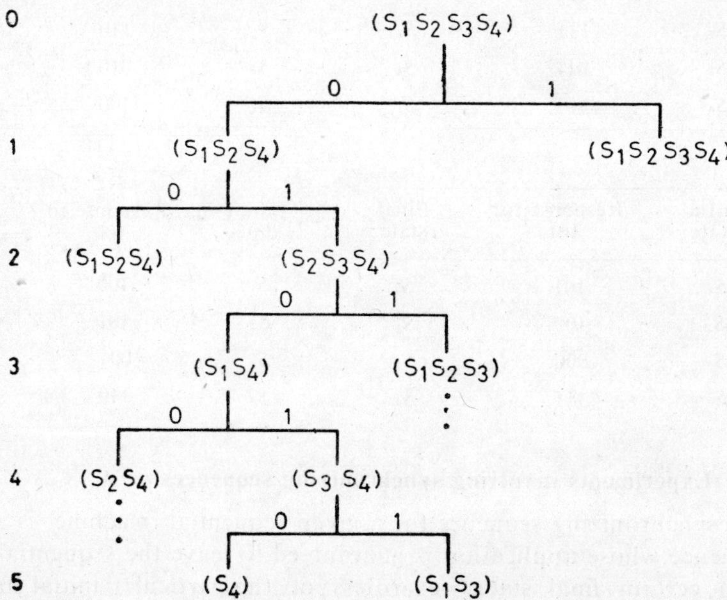


Fig. 2.3 Synchronizing tree for machine $M_3$.

single state ($S_4$) demonstrates the presence of a synchronizing sequence.

## 2.5 Characterizing sequences

A set of characterizing sequences for a sequential machine $M$ is a set of input sequences such that the set of output sequences produced in response to them by $M$ in any state $S_i$ is different from the set of output sequences produced by $M$ in any other state $S_j \neq S_i$. Thus a set of characterizing sequences serves to identify every state of the machine. Although there exist machines that do not have distinguishing sequences, all reduced machines have characterizing sequences. Sometimes a distinction is made between a characterizing set for a single state, and a set for a machine which is a characterizing set for every state of the machine. A set of characterizing sequences (or simply a characterizing set) for a state $S_i$ is a set of input sequences for which, with the machine in $S_i$ prior to the application of each element of the set, the set of responses uniquely identifies $S_i$. A characterizing set for a machine $M$ is a set of input sequences that is a characterizing set for every state of $M$. Let us take the example of a machine $M_4$ as shown in Table 2.6. Its corresponding uncertainty tree

Table 2.6 Machine $M_4$

| Present states | Next states and present outputs | |
|---|---|---|
| | $I_1 = 0$ | $I_2 = 1$ |
| $S_1$ | $S_1$, 0 | $S_2$, 0 |
| $S_2$ | $S_1$, 0 | $S_3$, 0 |
| $S_3$ | $S_1$, 0 | $S_4$, 0 |
| $S_4$ | $S_1$, 1 | $S_4$, 0 |

is depicted in Fig. 2.4. Here, unlike the earlier successor trees, the uncertainty partition vector pair is associated with the nodes in every level.

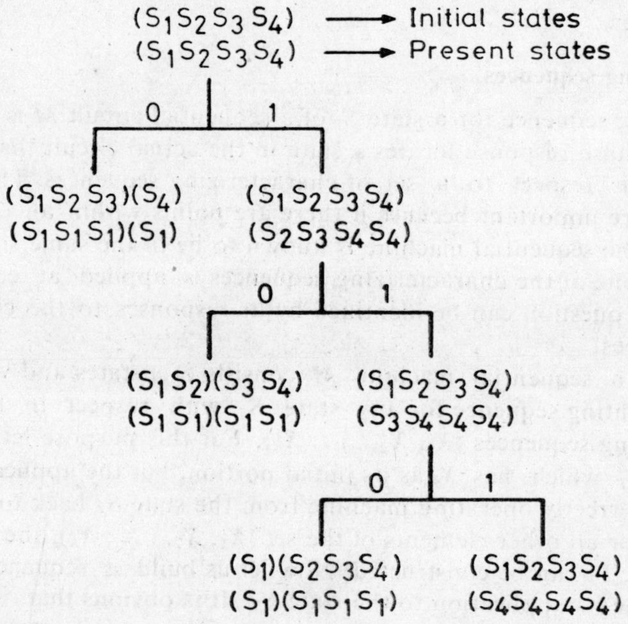


Fig. 2.4 Uncertainty tree for machine $M_4$.

A pair consisting of an initial state uncertainty partition and its associated present state uncertainty vector is known as uncertainty partition vector pair. The tree thus formed is infinite unless termination rules are specified. A branch at a node is terminated if either of the following occurs:

(i) The uncertainty partition vector pair associated with the node is identical (considering the ordering of states, blocks and components) to a pair associated with a node on a previous level.

(ii) A trivial present state uncertainty vector is associated with the node. Here the trivial present state uncertainty vector contains only homogeneous and singleton components.

From the above tree, a characterizing set for state $S_i$ is obtained by selecting sequences corresponding to a set of paths from the zero level node to nodes that have the property that their associated initial state uncertainty partitions may be multiplied to form a partition product that contains state $S_i$ as a singleton block. Similarly, a characterizing set for a machine is obtained by selecting a set of nodes for which the partition product is the trivial initial state uncertainty partition (the zero partition of the algebra). This comes from the fact that the partition product represents joint knowledge of initial state uncertainties associated with various input sequences.

From Fig. 2.4, it is seen that the set {0, 10} is a characterizing set for state $S_3$ of $M_4$, since the observation of the response set {0, 01} uniquely identifies the state $S_3$. Likewise, the set {0, 10, 110} is a characterizing set for machine $M_4$, yielding four different response sets corresponding to various states.

## 2.6 Locating sequences

A locating sequence for a state $S_i$ of a sequential circuit $M$ is an input sequence whose response locates a state in the actual circuit that behaves like $S_i$ with respect to a set of characterizing sequences. The locating sequences are important because if there are points within an experiment at which the sequential machine is known to be in the same state, and if a different one of the characterizing sequences is applied at each point, the state in question can be identified by its responses to the characterizing sequences.

Suppose a sequential machine $M$ consists of $n$ states and we want to design a locating sequence for any state $S_x$ with respect to the set of characterizing sequences $\{X_1, X_2, \ldots, X_k\}$. For this purpose let us take a sequence $Y_i$ which has $X_i$ as its initial portion, but the application of $Y_i$ takes the correctly operating machine from the state $S_x$ back to it again. Similarly, for all other elements of the set $\{X_1, X_2, \ldots, X_k)$, the sequences $Y_1, Y_2, \ldots, Y_k$ can be constructed. Now let us build a sequence $Y_1^{n+1}Y_2$, and examine its application to the state $S_x$. It is obvious that just before the application of $Y_2$, the circuit must have been in a state that responds to $X_1$ and $X_2$ like the state $S_x$ only.

Next let us consider the application of the sequence $(Y_1^{n+1}Y_2)^{n+1}Y_1^{n+1}Y_3$ to the state $S_x$. Assume the response of this sequence to the machine in state $S_x$ will be just like the correctly operating machine in state $S_x$. Though before the application of each of the $Y_2$ of the above sequence, the machine $M$ must have been in a state that responds to $X_1$ and $X_2$ like the state $S_x$, yet out of these at least $n$ states can be distinct. Thus, the state prior to the last application of $Y_2$ must be identical to the state prior to one of the other applications of $Y_2$, say the $j$-th one. Therefore, the application of $Y_2Y_1^{n+1}$ to the machine at a certain state which was just identified to be a state before the $j$-th application of $Y_2$, brings the machine to some state which responds to $X_1$ and $X_2$ as the state $S_x$ does. But this state in question is the state assumed just prior to the application of $Y_3$; therefore it is a state that responds to $X_1$, $X_2$, and $X_3$ as $S_x$ does.

Following the same line of argument, the correct response of the machine to the sequence, namely, $[(Y_1^{n+1}Y_2)^{n+1}Y_1^{n+1}Y_3]^{n+1}(Y_1^{n+1}Y_2)^{n+1}Y_1^{n+1}Y_4$ guarantees that the state of the circuit prior to the application of $Y_4$ is the one that responds correctly to the first four characterizing sequences. Repeating the process until $Y_k$ is reached yields a sequence that stands for the locating sequence for the state $S_x$ of the machine.

As an example to illustrate the building of locating sequence for a state in a sequential machine, let us take the machine $M_4$ as shown in Table 2.6 again. Now, as obtained earlier, the set $\{X_1 = 0,\ X_2 = 10,\ X_3 = 110\}$ constitutes a characterizing set for this machine. In order to design a locating sequence for state $S_2$ (say), it is first necessary to choose $Y_1$, $Y_2$, and $Y_3$. Applying $X_1$ to state $S_2$ leads to state $S_1$, from which the machine can be returned to state $S_2$ by applying a 1. Therefore $Y_1 = 01$. Similarly, $Y_2 = 101$, and $Y_3 = 1101$ can be determined. The sequence

$$[(01)^5 101]^5 (01)^5 1101$$

whose length is 79 constitutes a locating sequence for state $S_2$. In a similar way, the locating sequences for the other states can be designed.

## 2.7 Upper bounds on the length of the experiments

The lengths of the state identification experiments are easily calculated and these are discussed well in Gill [92], Kohavi [135], and Lee [143].

2.7.1 *A reduced q-state machine need not possess a preset distinguishing sequence, but if a distinguishing sequence exists it need not be longer than $(q - 1)q^q$ symbols.**

*Proof*: Suppose, the uncertainty vector at certain level of the distinguishing tree consists of $p$ components of sizes $k_1, k_2, \ldots, k_p$. The sum of the sizes of all the components must be equal to $q$, the number of states of the machine, or in other words, $k_1 + k_2 + \ldots + k_p = q$. Now take the numbers $k_1, k_2, \ldots, k_p$ as subsets in a partition $\eta$, such that $\eta = (k_1, k_2, \ldots, k_p)$. Thus $\eta$ defines the *size distribution* of the components in the uncertainty vector. The number of different uncertainty vectors with the same size distribution $\eta$ is equal to $q^{k_1}q^{k_2}\ldots q^{k_p} = q^q$. Now, let us take a path in the tree leading from the initial uncertainty vector to a trivial uncertainty vector. Let $U_1$ and $U_2$ be the two uncertainty vectors through this path having corresponding partitions $\eta_1$ and $\eta_2$. Obviously, if $U_2$ is a successor of $U_1$, then the size distribution of $U_2$ is either equal to that of $U_1$ or is a refinement of that of $U_1$, or, in other words, $\eta_1 \geqslant \eta_2$. As the initial uncertainty vector contains $q$ states, there are at most $(q - 1)$ refinements of partitions possible along the path leading to the

*The above bound is not necessarily the least upper bound.

distinguishing sequence. Therefore, if the length of this path is $L$, then $L$ is less than or equal to $(q - 1)q^q$ symbols.

2.7.2 *A preset homing sequence whose length is at most $(q - 1)^2$ exists for every reduced q-state machine M.*

*Proof*: Let the initial uncertainty be $(S_1, S_2, \ldots, S_q)$. Since $M$ is reduced, for every pair of states $(S_i, S_j)$ there exists an experiment of length $q - 1$ or shorter which distinguishes $S_i$ from $S_j$. Let us denote this experiment by $\lambda_1$. Starting with the initial-state uncertainty, the application of the sequence $\lambda_1$, which distinguishes between some pair of states in $M$, yields the $\lambda_1$-successor uncertainty vector, which contains at least two components. Next, select any two states in one component and apply the appropriate sequence $\lambda_2$, which distinguishes between them. The $\lambda_1\lambda_2$ uncertainty vector contains at least three components. Similarly, we obtain the $\lambda_1\lambda_2 \ldots \lambda_{q-1}$ successor vector, which consists of $q$ components, each of which contains only one state. Therefore, the sequence $\lambda_1\lambda_2 \ldots \lambda_{q-1}$ is a homing sequence. Since the length of each $\lambda_i$ experiment is at most $q - 1$, the total length of the homing sequence cannot exceed

$$\underbrace{(q - 1) + (q - 1) + \ldots + (q - 1)}_{(q-1) \text{ times}} = (q - 1)^2$$

2.7.3 *If a synchronizing sequence for a q-state machine M exists then its length is at most $(q - 1)^2q/2$.*

*Proof*: Let the initial uncertainty be $(S_1, S_2, \ldots, S_q)$, and let us take any two states $(S_i, S_j)$. Since the machine $M$ is known to possess a synchronizing sequence, let us apply a sequence $\lambda_1$ to the machine which takes the above mentioned states $(S_i, S_j)$ to a state, say $S_k$. The length of the sequence $\lambda_1$ is atmost $(q - 1)\, q/2$, since the longest path for the synchronization of $(S_i, S_j)$ is through all possible pairs of states, i.e., $(S_1, S_2), (S_1, S_3), \ldots, (S_{q-1}, S_q)$. Thus the state $S_k$ becomes $\lambda_1$-successor of $(S_i, S_j)$. Next, we select a state $S_m$ from the resultant uncertainty and determine a sequence $\lambda_2$ which takes $(S_k, S_m)$ into some state, say $S_n$. The length of $\lambda_2$ is atmost $(q - 1)q/2$. Similarly, it is possible to find the sequences $\lambda_3, \lambda_4, \ldots, \lambda_{q-1}$, which, when concatenated, yield the synchronizing sequence $\lambda_1\lambda_2 \ldots \lambda_{q-1}$ whose length is at most $(q - 1)^2q/2$.

2.7.4 *A characterizing set of sequences exists for every reduced q-state machine, and that such a set consists of, in worst case, $q - 1$ elements of length of at most $(q - 1)$ symbols.*

*Proof*: If the machine is reduced, it is always possible to characterize its states by their responses to an appropriately chosen set of input sequences. The number of sequences in the set need never exceed $q - 1$, and no sequence need exceed $q - 1$ in length. To prove the same, note that if no two states in the transition table are equivalent, then any two given states can be distinguished by an input sequence of length at most $(q-1)$. Arbitrarily select two states and design a sequence that distinguishes between them.

The set of states may now be partitioned into at least two classes according to their responses to this particular sequence. Arbitrarily, again take two members of some class and design an input sequence that distinguishes between them. The states may now be partitioned into at least three classes according to their responses to the two sequences designed so far. Continuing in the above manner, it yields a set of atmost $q-1$ sequences that distinguishes among all the states, with each individual sequence having length atmost $(q-1)$ symbols.

## 2.8 Simple I/O sequence

Let $\pi$ be an I/O sequence generated by the state $S_u$ of a sequential machine $M$. $\pi$ is said to be a simple I/O sequence of $M$ if no state other than $S_u$ of $M$ can generate the said I/O sequence.

Given a machine $M$, a simple I/O sequence distinguishes one state from all other states of $M$, while a DS distinguishes each state from every other state of the machine. If $\pi$ is verified to be a simple I/O sequence for the machine under test, then the occurrence of the simple I/O sequence $\pi$ at two different points of the experiment establishes that the machine must have been in the same final state after each occurrence of $\pi$.

## 2.9 Compound distinguishing sequences

A compound DS of a machine $M$ is a set of input sequences $X = \{X_1, X_2, \ldots, X_r\}$ having a common prefix $p_k$ but otherwise being unrestricted, with the following properties: (i) For each set $P$ of $p_k$-equivalent states of $M$, there is a sequence $X_i$ in $X$ that distinguishes the states in $P$. (ii) The deletion of any sequence from the set $X$ will leave some set of $p_k$-equivalent states not distinguished by any $X_i \in X$.

A DS can be considered as a special compound DS, $X = \{X_1, \ldots, X_r\}$ in which $X_i = X_j$ for all $i, j$, whereas, in general, only a certain prefix $p_k$ of each $X_i$ is required to be equal for a compound DS.

## 2.10 Resolving sequences

A machine $M$ with $q$ states is said to be $(q/2)$-resolvable if there exists an input sequence $X$ such that $M$ produces at least $\lceil (q+1)/2 \rceil$ different output sequences when $X$ is applied to $M$ in any one of the initial states belonging to the state alphabet $(S_1, S_2, \ldots, S_q)$, where $\lceil r \rceil$ denotes the smallest integer greater than or equal to $r$. Here $X$ stands for the resolving sequence of the machine $M$.

An input sequence $X$ induces a partition $p_x$ on the set of states of $M$ into blocks of the $X$-equivalent states where two states are in the same block, if and only if, their output responses to $X$ are identical.

## 2.11 Compound resolving sequence

A compound resolving sequence (CRS) of $M$ is a set of input sequences $[XY_1, \ldots, XY_r]$ with a common prefix $X$ such that the following statements are true:

(i) The prefix $X$ partitions the set of states of $M$ into $r$ $X$-equivalent classes $S_1, \ldots, S_r$.

(ii) If $M$ responds to $XY_i$ with $m_i$ different outputs when starting from the states in $S_i$, then

$$\sum_{i=1}^{r} m_i \geqslant \lceil (q+1)/2 \rceil,$$

where $q$ is the number of states of the machine.

## 2.12 Variable-length distinguishing sequences

A variable-length distinguishing sequence (VLDS) is a preset distinguishing sequence $X_0$ such that, if the machine is started in an unknown state, the output response of the machine to some prefix of $X_0$ will identify the initial state.

The length of the required prefix of $X_0$ will identify the initial state. Consider, for example, the sequential machine as shown in Table 2.1, whose response to the distinguishing sequence 111 is as shown below from Table 2.4.

| Initial states | Responses to 111 | Final states |
|---|---|---|
| $S_1$ | 110 | $S_2$ |
| $S_2$ | 111 | $S_3$ |
| $S_3$ | 011 | $S_4$ |
| $S_4$ | 101 | $S_1$ |

It is evident that if the response of the machine to the first input symbol is 0, the initial state must have been $S_3$ and the distinguishing experiment may be terminated at this stage. On the other hand, if the response is 1, the initial state could have been either $S_1$, $S_2$, and $S_4$. Hence the experiment must continue, and the machine is supplied with the second input 1. If the machine's response is now 0, the initial state must have been $S_4$, and the distinguishing experiment may be terminated. If, however, the response is 1, the uncertainty regarding the initial state is $(S_1S_2)$, and a third input 1 must be applied to the machine. Thus, for the machine in question, and for the distinguishing sequence 111, the shortest distinguishing prefix (SDP) for state $S_3$ is 1; for state $S_4$ it is 11; for states $S_1$ and $S_2$ it is 111.

## 2.13 Definitely diagnosable machine

A machine $M$ is defined as a definitely diagnosable machine (DD machine) of order $l$ if $l$ is the least integer, so that every sequence of length $l$ is a distinguishing sequence for $M$. Here the integer $l$ is called the order of diagnosability. In other words, a machine becomes definitely diagnosable if every node in level $l$ of the distinguishing tree is associated with a trivial uncertainty vector. A distinguishing tree can thus serve as a tool for recognizing definitely diagnosable machines.

Whether a sequential machine possesses any DS or not, it can be converted to a definitely diagnosable machine having specified distinguishing sequences by augmenting the machine with some extra output variables. A detailed discussion is given in Chapter IV.

## 2.14 Counter-cycle machine

A sequential machine whose transition graph has a counter-cycle $C_I$ for some input symbol $I$ is said to be counter-cycle machine (CC machine).

A counter-cycle $C_I$ of an $n$-state sequential machine is an alternating sequence of states $S_i$ $(1 \leqslant i \leqslant n)$ and input symbol $I$ such that

$$S_1 \xrightarrow{I} S_2 \xrightarrow{I} S_3 \rightarrow \ldots \xrightarrow{I} S_n \xrightarrow{I} S_1$$

is satisfied in the transition graph. If the next states and output functions are given by the mappings $f$ and $g$ respectively, then counter-cycle with input $I$ is defined as follows:

$$f(S_i, I) = S_{i+1} \text{ and } g(S_i, I) = 0^* \quad (1 \leqslant i \leqslant n-1)$$

$$f(S_n, I) = S_1 \quad \text{and } g(S_n, I) = 1^*$$

A detailed discussion is included in Chapters IV and V.

## 2.15 Output-observable machine

A sequential machine $M$ is called $k_1, k_2, \ldots, k_p$-output observable with respect to the output function $z_1 \times z_2 \times \ldots \times z_p$ and a partition $\pi$ if the following conditions are satisfied, where each $k_i$ is a nonnegative integer.

*Condition 1*: The knowledge of the present state of $M$ is sufficient to uniquely determine the succeeding output sequence of length $k_j$ observed at the output function $z_j$ for every $j$ $(j = 1, 2, \ldots, p)$.

*Condition 2*: Let $\mu_{ij}$ be the output sequence of length $k_j$ observed at $z_j$ when the initial state is $S_i$. Then $S_i \underset{\pi}{\sim} S_j$†, if and only if, $(\mu_{i1}, \ldots, \mu_{ip}) = (\mu_{j1}, \ldots, \mu_{jp})$, for all $S_i$ and $S_j \in S$. When $\pi$ is the zero partition, then only $M$ is called output observable. For example, the sequential machine $M_5$ shown in Table 2.7 is 1-output observable with respect to $z_1$ and $\pi_1 = \{\overline{S_1S_2}; \overline{S_3S_4S_5}\}$. A sequential machine $M_6$ shown in Table 2.8 is 1, 2-

---

*The output symbols need not be restricted to 0 and 1 if only they are distinct from each other. Without loss of generality, the symbols 0 and 1 are used as shown above.

†A relation $\underset{\pi}{\sim}$ on $S$, the state alphabet of the machine, corresponding to a partition $\pi$ is a relation such that $S_i \underset{\pi}{\sim} S_j$ for $S_i, S_j \in S$ if and only if $S_i$ and $S_j$ belong to the same block of $\pi$.

output observable with respect to $z_1 \times z_2$ and the zero partition, and thus the machine $M_6$ is output observable by definition.

Table 2.7 Machine $M_5$

| Present states | Next states and present outputs across $z_1$ | |
|---|---|---|
| | $I_1 = 0$ | $I_2 = 1$ |
| $S_1$ | $S_4$, 0 | $S_2$, 0 |
| $S_2$ | $S_3$, 0 | $S_2$, 0 |
| $S_3$ | $S_4$, 1 | $S_3$, 1 |
| $S_4$ | $S_5$, 1 | $S_5$, 1 |
| $S_5$ | $S_5$, 1 | $S_1$, 1 |

Table 2.8 Machine $M_6$

| Present states | Next states and present outputs across $z_1$ and $z_2$ | |
|---|---|---|
| | $I_1 = 0$ | $I_2 = 1$ |
| $S_1$ | $S_4$, 01 | $S_2$, 01 |
| $S_2$ | $S_3$, 00 | $S_2$, 00 |
| $S_3$ | $S_4$, 10 | $S_5$, 10 |
| $S_4$ | $S_5$, 10 | $S_5$, 10 |
| $S_5$ | $S_5$, 11 | $S_1$, 11 |

## 2.16 Distinguishing set

The concept of D-set was introduced by R. T. Boute [32]. A sequential machine $M$ is defined as usual by the quintuple $(I, S, O, f, g)$ where $I$, $S$, and $O$ are, respectively, the input, state and output alphabets of the machine, whereas $f$ and $g$ denote the next state and the output mappings, respectively. Let $X^*$ denote a set of finite input sequences together with the empty or null sequence $\lambda$.

A distinguishing set ($D$-set) $D$ for a machine $M$ is a set of input sequences $\bar{x}_i$, one for each state $S_i \in S$, such that for every pair $S_iS_j$ of different states we can write $\bar{x}_i = \overline{u_{ij}}\ \overline{y_{ij}}$ and $\bar{x}_j = \overline{u_{ij}}\ \overline{z_{ij}}$ where the sequences $\overline{u_{ij}}, \overline{y_{ij}}, \overline{z_{ij}} \in X^*$ are such that $g(S_i, \overline{u_{ij}}) \neq g(S_j, \overline{u_{ij}})$. {Here $g(S_i, \bar{x})$ denotes the sequence of output symbols produced when the machine is started in state $S_i$ and the input sequence $\bar{x}$ is applied}.

EXAMPLE: For the machine $M_7$ shown in Table 2.9, the $D$-set is given in Table 2.10.

Table 2.9 Machine $M_7$

| Present states | Next states and present outputs | |
|---|---|---|
| | $I = 0$ | $I = 1$ |
| $S_1$ | $S_2$, 0 | $S_4$, 1 |
| $S_2$ | $S_3$, 1 | $S_4$, 0 |
| $S_3$ | $S_2$, 0 | $S_1$, 1 |
| $S_4$ | $S_1$, 0 | $S_2$, 0 |

Table 2.10 A distinguishing set for $M_7$

| States $q$ | Inputs $\bar{x}_i \in D$ | Responses to $\bar{x}_i$ | Final states |
|---|---|---|---|
| $S_1$ | 11 | 10 | $S_2$ |
| $S_2$ | 10 | 00 | $S_1$ |
| $S_3$ | 11 | 11 | $S_4$ |
| $S_4$ | 10 | 01 | $S_3$ |

It is easy to verify that these sequences can be decomposed in the fashion implied by the definition, e.g., if $S_i = S_1$ and $S_j = S_2$, then $\overline{u_{ij}} = 1$, $\overline{y_{ij}} = 1$ and $\overline{z_{ij}} = 0$, while for $S_i = S_3$ and $S_j = S_1$, we have $\overline{u_{ij}} = 11$, $\overline{y_{ij}} = \lambda$, $\overline{z_{ij}} = \lambda$.

## 2.17 Maximum flow table

A special type of flow table known as the maximum flow table is sometimes encountered in sequential machine identification problem [55, 205, 206]. If any input/output sequence of a given sequential machine is known then the said input/output sequence can be used for the construction of maximum flow table by assuming that the machine goes to a new state every time an input sequence is applied. For a given input/output sequence of length $L$, the maximum flow table has $L + 1$ rows and $p$ columns, $p$ being the number of input symbols; each row of the flow table except the last has only one specified next state entry in a certain column, whereas the last row has all unspecified next state entries in all the columns.

EXAMPLE : Suppose an input-output sequence sufficient to identify a four state Moore machine $M$ is given as

$$X_i = 0111 \quad 0100 \quad 0101 \quad 11$$

$$Z_k = 0011 \quad 0010 \quad 1000 \quad 11$$

Let us consider a subsequence of the given input-output sequence of length 9 as shown below:

$$X_{ia} = 0111 \quad 0100 \quad 0$$

$$Z_{ka} = 0011 \quad 0010 \quad 1$$

We can construct a maximum flow table $F'$ with respect to the input/output sequence $X_{ia}/Z_{ka}$. The maximum flow table $F'$ is shown in Table 2.11.

Table 2.11 Maximum flow table $F'$

| Present states | Next states | | Present outputs |
|---|---|---|---|
| | $I_1 = 0$ | $I_2 = 1$ | |
| $S_1$ | $S_2$ | — | — |
| $S_2$ | — | $S_3$ | 0 |
| $S_3$ | — | $S_4$ | 0 |
| $S_4$ | — | $S_5$ | 1 |
| $S_5$ | $S_6$ | — | 1 |
| $S_6$ | — | $S_7$ | 0 |
| $S_7$ | $S_8$ | — | 0 |
| $S_8$ | $S_9$ | — | 1 |
| $S_9$ | $S_{10}$ | — | 0 |
| $S_{10}$ | — | — | 1 |

## 2.18 Direct sum table

Sometimes it is necessary to combine the state table descriptions of two or more machines into a single state. This is done simply by adding the table of one machine to the bottom of the table of the other. Of course before doing the same, it is made sure that no state of one machine has the same name as a state in the other. Suppose it is necessary to combine the state table descriptions of the machine $M_1$ and $M_2$ shown in Table 2.12 (a) and (b) respectively. The resulting table $M_1 + M_2$ thus obtained in this way represents the direct sum table of the two given machines $M_1$ and $M_2$ and is shown in Table 2.13.

Table 2.12

(a) Machine $M_1$

| Present states | Next states and present outputs | |
|---|---|---|
| | $I = 0$ | $I = 1$ |
| $S_1$ | $S_2$, 0 | $S_1$, 0 |
| $S_2$ | $S_1$, 0 | $S_3$, 1 |
| $S_3$ | $S_3$, 1 | $S_1$, 0 |

(b) Machine $M_2$.

| Present states | Next states and present output | |
|---|---|---|
| | $I = 0$ | $I = 1$ |
| $S_4$ | $S_5$, 0 | $S_4$, 0 |
| $S_5$ | $S_4$, 0 | $S_7$, 1 |
| $S_6$ | $S_5$, 0 | $S_6$, 1 |
| $S_7$ | $S_7$, 1 | $S_4$, 0 |

Table 2.13 Machine $M_1 + M_2$

| Present states | Next states and present outputs | |
|---|---|---|
| | $I = 0$ | $I = 1$ |
| $S_1$ | $S_2$, 0 | $S_1$, 0 |
| $S_2$ | $S_1$, 0 | $S_3$, 1 |
| $S_3$ | $S_3$, 1 | $S_1$, 0 |
| $S_4$ | $S_5$, 0 | $S_4$, 0 |
| $S_5$ | $S_4$, 0 | $S_7$, 1 |
| $S_6$ | $S_5$, 0 | $S_6$, 1 |
| $S_7$ | $S_7$, 1 | $S_4$, 0 |

The equivalent states in two different machines can be found by applying the partitioning process to the direct sum of the two machines. For example, the direct-sum machine of Table 2.13 yields the following partitions:

$$P_1 = (S_1, S_4)(S_2, S_5, S_6)(S_3, S_7)$$

$$P_2 = (S_1, S_4)(S_2, S_5)(S_6)(S_3, S_7)$$

$$P_3 = P_2$$

These partitions show that state $S_1$ in machine $M_1$ is equivalent to state $S_4$ in machine $M_2$, state $S_2$ in machine $M_1$ is equivalent to state $S_5$ in machine $M_2$ and state $S_3$ in machine $M_1$ is equivalent to state $S_7$ in machine $M_2$; whereas no state in machine $M_1$ (or in $M_2$) is equivalent to state $S_6$ of machine $M_2$.

Now by definition, if for every state in machine $M_1$ there exists at least one equivalent state in $M_2$, then $M_2$ is said to contain $M_1$. If for every state of $M_1$ there exists at least one equivalent state of $M_2$ and vice versa, then the machines $M_1$ and $M_2$ are said to be equivalent. In the above stated example, the equivalence partition of the direct sum machine is $(S_1, S_4)(S_2, S_5)(S_6)(S_3, S_7)$ and since each block contains one state from the machine $M_2$, therefore $M_2$ contains $M_1$. As the block $(S_6)$ does not contain any state from the machine $M_1$, therefore, $M_1$ does not contain $M_2$ and the two machines are not equivalent. If $M_2$ starts from the initial state $S_6$, there is no way of finding an initial state of $M_1$ so as to make the terminal behaviour of the two machines identical. Further, if we do not know even the starting state of the $M_1$, there is some initial state of $M_2$ for which the two machines will have identical terminal behaviour.

## 2.19 $I_i$-mergeable machine

Let a sequential machine be defined by a quintuple $M = (I, S, O, f, g)$ as stated in Section 2.16. The machine $M$ is said to be $I_i$-mergeable, if and only if, for each input symbol $I_i \in I$, there exist at least two states

$S_i, S_j \in S$, such that $f(S_i, I_i) = f(S_j, I_i)$ and $g(S_i, I_i) = g(S_j, I_i)$. Otherwise, the machine is called non-$I_i$-mergeable. The Table 2.14 provides an illustration of an $I_i$-mergeable machine. Here the states $S_1$ and $S_2$ are merged into the state $S_3$ under the input symbol 0 resulting in an output symbol 0 in both the cases, while the states $S_2$ and $S_3$ are merged into the state $S_2$ under the input symbol 1 producing an output symbol 1 in both the cases. The sequential machine which are $I_i$-mergeable do not possess any diagnosing sequences. This is because every uncertainty vector in the first level of the successor tree of $I_i$-mergeable machine is homogeneous uncertainty vector. Hence an $I_i$-mergeable machine does not possess a distinguishing sequence because there results no trivial uncertainty vector.

Table 2.14 $I_i$-mergeable machine

| Present states | Next states and present outputs | |
|---|---|---|
| | $I = 0$ | $I = 1$ |
| $S_1$ | $S_3$, 0 | $S_1$, 0 |
| $S_2$ | $S_3$, 0 | $S_2$, 1 |
| $S_3$ | $S_1$, 1 | $S_2$, 1 |

## 2.20 Easily testable machine

An easily testable sequential machine as was originally defined by Fujiwaral *et al.* [85] is one for which a short preset checking experiment can readily be designed following a simple algorithm. This is possible because such a machine possesses a short distinguishing sequence, a short synchronizing sequence and short transfer sequences. For details, the interested reader may refer to the Section 4.6 of this book. Of course, later on various easily testable machine models are discussed in Sections 5.5, 5.6 and 5.7 respectively.

# Chapter 3

# Designing Checking Experiments for Sequential Machines–A Comprehensive Review

## 3.1 Introduction

In the preceding chapter, the various state identification experiments and other basic terminologies were discussed. In this chapter, some existing ideas pertaining to the problem of identifying an unknown finite state machine and that of designing its checking sequences will be considered. The machine identification experiments are basically concerned with the problem of determining whether or not a given $q$-state sequential machine is distinguishable from all other $q$-state machines. On the other hand, an experiment designed to take a sequential machine through all its state transitions, in a way such that a definite conclusion can be reached as to whether or not the machine is operating correctly, is said to be a checking experiment or a fault-detection experiment.

The machine identification problem is a general problem involving terminal measurements where the machine $M$ of which the behaviour is to be determined is considered to be given to the experimenter as a "black box". An input sequence is applied to $M$, and the corresponding output sequence is recorded. The experimenter's objective is to find from this input-output sequences the state table that represents the given machine. This problem has a solution, in general, if the machine being tested is *reduced*, *strongly connected** and an upper bound on the number of its states is known. The designing of fault-detection experiments may be considered as the reverse problem of machine identification. An experimenter in this case is supplied with a machine and its state table, and his job in this particular situation is to determine, from the terminal measurements alone, whether the given state table accurately describes the behavior of the machine, or, in other words, if the actual machine

*A machine is said to be a *strongly connected* machine if and only if it is possible to reach any state $S_j$ from any other state $S_i$ or, in other words, if and only if every state is accessible from every other state.

is *isomorphic*† to the one described by the given state table.

In dealing with the case of machine identification, a question that naturally arises at this point is whether it is possible to find a state table description, if no information at all is available about the state table behavior of a given machine. Posed in such a general form, this problem has no solution. Given any machine and any finite length $k$, it is always possible to construct another machine that behaves exactly like the given one for all experiments of length $k$ or less, but behaves differently than the given machine for some experiments of length greater than $k$. Thus no such finite experiment can ever completely determine the state table of a particular machine. However, if the machine is known to have at most $q$-states, the problem of finding its state table reduces to one of determining which of a finite number of state tables describes the behaviour of the machine. Of course, it will be assumed here that there are no equivalences that involve the states from the various tables. To achieve this, a direct sum of the various tables can be formed, and either a preset or an adaptive homing sequence can be designed for the direct sum machine. It is always possible to find a quasi-homing experiment that describes the final state of the sum machine to within a set of equivalent states. As it has been stated above that the two states could be equivalent if they belong to the same table only, this homing experiment thus provides enough information to decide upon which of the tables contains the final state.

When a machine is made available to the investigator in physical form, it may be possible to place an upper bound on the number of its states, perhaps by counting the number of elements in the circuit. It is then possible to determine what the behavior of the given machine will be in future. For this purpose, first the direct sum of all the possible state tables having $q$ or fewer states can be formed, and a quasi-homing experiment for the sum table can also be constructed for the purpose of applying it to the given machine. This experiment will leave the machine in a known set of equivalent states from the sum table. Any one component of the sum table that contains one of these equivalent states can thus be located out as to be a table $T$ which contains a state $S$. The future behavior of the given machine will then be the same as that of the machine described by this table $T$ being set at the state $S$.

A similar approach can be taken with the problem of determining if or not a given machine is operating correctly. Unfortunately, the finite checking experiments do not exist unless some restrictions are imposed on the original correct machine and the malfunctions that the machine may experience. At first, the correctly operating circuit must correspond to that of a strongly connected machine (although the faulty circuit need

†The terminal behaviour of a machine is not at all influenced by the names given to its states. Technically, two tables or machines that can be obtained from one another by the remaining states are said to be *isomorphic*.

not). Secondly, there should be a finite number of malfunctions that must occur in the circuit, and also the effects of these malfunctions should be known. E. F. Moore [157] in 1956 showed in his classic paper that if both of these restrictions were met, in principle at least it is possible to design a checking experiment for the given circuit. His experiment not only determines whether or not one of the allowed set of malfunctions has occurred, but also determines specifically which of these malfunctions has occurred. As long as the number of states in a machine can be bounded, it is theoretically possible to design experiments that will specify the machine's state table, or indicate whether or not the machine is functioning correctly. This can be done, knowing beforehand that the number of states of the actual machine is at most $q$, by forming the direct sum of all the possible tables that have $q$ or fewer states, and then finding a quasi-homing experiment for it. As stated earlier, the application of this homing sequence to the machine at hand for the experimentation will reveal which set of the equivalent states from the sum table contains the final state of the machine. If none of the states in this set belongs to the original state table, then the table no longer describes the behavior of the machine. If at least one of the states in the set belongs to the original state table, then the table accurately describes the future behavior of the machine. This, of course, remains valid so long as we exclude all possible future malfunctions that might occur.

The method just outlined has one big disadvantage of requiring an extraordinary amount of work to develop an experiment that checks against a large number of malfunctions. If an experiment is to check against all the malfunctions of a circuit having at most $q$ states, then it is necessary to consider more than $q^{2q}$ different state tables. For example, suppose we would like to determine if a certain machine is operating in accordance with the state table as given in Table 3.1, or if it is in reality described by some other 3-state table. To know the exact answer to such a problem, we should have to list all the nonequivalent tables having three or fewer states. But as there are more than 7000 such tables, it is quite impossible to accomplish this job by hand. Even a modern digital computer is of little help for handling bigger tables, with large numbers of states.

Table 3.1 Machine $M_1$

| Present states | Next states and present outputs | |
|---|---|---|
| | $I_1 = 0$ | $I_2 = 1$ |
| $S_1$ | $S_2$, 1 | $S_1$, 0 |
| $S_2$ | $S_3$, 0 | $S_1$, 1 |
| $S_3$ | $S_1$, 0 | $S_2$, 0 |

## *3.2 Transition-checking approach as a basic tool of fault detection

F. C. Hennie [113] describes a completely different but novel approach for designing checking experiments for sequential machines. Since it involves only the state table of the correctly operating circuit, not only his method enjoys the advantage of being easy to implement but also it remains capable of protecting against a large number of malfunctions. It would be appropriate here to consider an analytical example to demonstrate the ideas behind the formation of this type of checking experiments. Suppose an input sequence 1100 0001 0010 10 is applied to a certain sequential machine $M_1^*$ having three states only, with the hope of determining whether or not the machine is described by the state table as given earlier in Table 3.1. If $M_1^*$ is in fact described by this table, it will certainly be in state $S_1$ after the application of first two 1's in the sequence, and from that point onwards it will produce the output sequence 1001 0000 0111 in response to the said input sequence. If $M_1^*$ does not produce this particular output sequence, clearly then it is not described by the said state table. On the other hand, let us try to draw some conclusion about the state table description if the machine does in fact produce the output sequence $Z$ in response to the input sequence $X$ as shown below.

$$X: \; 0 \; 0 \; 0 \; 0 \; 0 \; 1 \; 0 \; 0 \; 1 \; 0 \; 1 \; 0$$

$$Z: \; 1 \; 0 \; 0 \; 1 \; 0 \; 0 \; 0 \; 0 \; 0 \; 1 \; 1 \; 1$$

It can be noted from the above input-output sequence that three different responses to the input sequence 00 have been produced throughout the course of the experiment. As the sequential machine is known to have three states only, we can infer that each state has been visited at least once during the course of the experiment. Furthermore, as each state produces a different output sequence when the input sequence 00 is applied, the sequence of input symbols 00 is a distinguishing sequence for the circuit. Let us use the names $S_1'$, $S_2'$, and $S_3'$ to denote the state in which $M_1^*$ responds to 00 by producing 10, 00 and 01, respectively. Also it becomes evident that whenever $M_1^*$ produces identical responses to 00 at two different points in the experiment, it must have been in the same internal state at each of these points. Thus the states that the given machine assumes at certain points in the course of the experiment can be identified as follows:

| $X$: | | 0 | | 0 | | 0 | | 0 | | 0 | | 1 | | 0 | | 0 | | 1 | | 0 | | 1 | | 0 |
|---|---|---|---|---|---|---|---|---|---|---|---|---|---|---|---|---|---|---|---|---|---|---|---|---|
| | $S_1'$ | | $S_2'$ | | $S_3'$ | | $S_1'$ | | | | | | $S_2'$ | | | | | | $S_1'$ | | | | $S_1'$ | |
| $Z$: | | 1 | | 0 | | 0 | | 1 | | 0 | | 0 | | 0 | | 0 | | 0 | | 1 | | 1 | | 1 |

*The example of the checking experiment/machine identification incorporated in section 3.2 (with reference to the Tables 3.1, 3.2, 3.3 and 3.3′ of this book) is produced with the permission of M/s John Wiley & Sons, Inc., publishers. This is taken from the book 'Finite-state Models for Logical Machines', authored by F. C. Hennie, copyright © 1968, John Wiley & Sons, Inc. "Translated by the permission of John Wiley & Sons, Inc. All rights reserved."

Note that since $S_1'$ is the only state that produces an output of 1 in response to the input of 0, it can be identified on that basis alone. It is obvious from the above input-output sequence that an input 0 causes the machine to go from $S_1'$ to $S_2'$, from $S_2'$ to $S_3'$, and from $S_3'$ to $S_1'$. These known transitions thus worked out enable us to fill the states assumed by the machine at the remaining points in the experiment as shown below:

| $X$: | 0 | 0 | 0 | 0 | 0 | 1 | 0 | 0 | 1 | 0 | 1 | 0 |
|---|---|---|---|---|---|---|---|---|---|---|---|---|
| | $S_1'$ | $S_2'$ | $S_3'$ | $S_1'$ | $S_2'$ | $S_3'$ | $S_2'$ | $S_3'$ | $S_1'$ | $S_1'$ | $S_2'$ | $S_1'$ |
| $Z$: | 1 | 0 | 0 | 1 | 0 | 0 | 0 | 0 | 0 | 1 | 1 | 1 |

A completed state table of the sequential machine $M_1^*$ can thus be obtained from the above input-output sequence, and is shown in Table 3.2.

Table 3.2 Machine $M_1^*$

| Present states | Next states and present outputs | |
|---|---|---|
| | $I_1 = 0$ | $I_2 = 1$ |
| $S_1'$ | $S_2'$, 1 | $S_1'$, 0 |
| $S_2'$ | $S_3'$, 0 | $S_1'$, 1 |
| $S_3'$ | $S_1'$, 0 | $S_2'$, 0 |

Though it may appear for the moment that $X$ and $Z$ were carefully chosen to provide exactly the desired information in a convenient form, yet some definite conclusions can be drawn from the experiment just described. Evidently, if some 3-state sequential machine responds to an input sequence $X$ by producing the output sequence $Z$, then its state table must be isomorphic to that of Table 3.2. Therefore, with the restriction that no malfunction increases the number of states, the sequence $X$ stands here for the checking sequence of the machine of Table 3.2 with $S_1'$ as the starting initial state. Since the machine of Table 3.1 is isomorphic to that of Table 3.2, obviously if the machine $M_1$ operates correctly and $X$ is applied to it at the state $S_1$, it will produce $Z$ as an output response. Conversely, it can also be stated that if the said machine actually produces the output sequence $Z$ in response to the input sequence $X$, it must have been operating correctly, and it has also started at the state $S_1$. Indeed, it is required to ensure at this point that if the circuit is operating correctly, it should be in state $S_1$ when $X$ is applied to it. In this particular example, it is also surely true because the synchronizing sequence 11 is guaranteed to leave the machine in the state $S_1$ when $X$ will be applied to it. For checking the given machine of Table 3.1 for failures, simply it is required to apply the combined input sequence 1100 0001 0010 10 to the machine and to note down the output sequence it produces, ignoring only the first two output symbols. If the output sequence thus observed is $Z$, then we may conclude that the machine is either operating correctly, or else has more than three states. But the machine is definitely faulty if the output sequence differs from $Z$.

It follows from the above discussion that the two basic properties of the aforementioned input-output sequence have made possible to get it utilized as a checking sequence for the machines under consideration. At first, it enables us to conclude that at certain points in the experiment, the sequential machine was in some specific states. Secondly, it also signifies that at certain points in the experiment, a single input symbol caused a transition between some states that can also be specified very well. The states following the transition can be identified by the response to the distinguishing sequence. The state preceding the transition can be identified if it is reached from some already identified state by a certain input sequence, and by virtue of another part of this experiment, if we have already identified the said state which can be reached in that way.

The checking experiment thus described may be divided into two parts. The first part of each such experiment consists of the application of an input sequence that will take the machine into some predesignated starting state, provided no malfunction will come in the way. If a synchronizing sequence exists for a given state table, then this part of the experiment can be preset. Otherwise, it will be necessary to apply a homing sequence (all reduced machines possess such sequences) to determine the new final state of the circuit, and then following it with an additional input sequence in order to transfer the machine at the desired starting state.

The second and the more important part of the experiment is designed to distinguish between a correctly operating circuit initially in the designated starting state, and the circuit that has suffered one of a specified class of malfunctions. This last part of the experiment, usually called a checking sequence, is designed in two steps. First, a locating sequence is constructed for each state and included somewhere within the checking sequence. Then the same sequence is extended to cover the transition checking of each individual transition in the given state table.

If the state table of a sequential machine is reduced, and strongly connected, and possesses a distinguishing sequence, the design of a checking sequence becomes quite easy, if and only if, no malfunction will cause an increase in the number of states of the machine. The distinguishing sequence itself serves here as a locating sequence for each state. Each transition is checked by manoeuvering the circuit into a state that has been previously identified by means of a distinguishing sequence, then applying the particular input symbol in question, and finally applying the distinguishing sequence to identify the state to which the transition leads.

The designing of a checking sequence becomes complicated if the given table does not possess a distinguishing sequence. The locating sequences are then formed by compounding the various characterizing sequences, and the checking of a single transition requires traversing that transition once for each element of the characterizing set of the machine.

The above-enumerated second part of the checking experiment is a preset one, and in case the circuit fails to respond correctly, there is also no basis for altering the input sequence. As soon as the circuit fails to respond correctly, it is known to be faulty, and altering the input sequence in such a case provides no further information. *This experiment is not concerned about finding what is actually wrong with the circuit, but only about whether or not something has gone wrong.*

A general synthesis procedure for designing checking experiments for sequential machines possessing distinguishing sequences was also provided by Hennie [113]. Though his general procedure does not always lead to the shortest possible checking sequence, yet it does provide an upper bound on the length of the checking sequence in terms of the length of the distinguishing sequence. For a $q$-state sequential machine having the state alphabet $\{S_1, S_2, \ldots, S_q\}$ with a distinguishing sequence $X_0$ of length $L_{ds}$, this general synthesis procedure goes as follows.

The first part of the checking sequence is applied with the circuit being in state $S_1$, and has the form

$$X_0T(Q_1, S_2)X_0T(Q_2, S_3)X_0T(Q_3, S_4) \ldots X_0T(Q_q, S_1)X_0$$

where the application of $X_0$ to the machine in any state $S_i$ leaves it to a state $Q_i$ (any two states $Q_i$ and $Q_j$ may be identical even though $i \neq j$), and $T(S_i, S_j)$ denotes some input sequence that takes the circuit from state $S_i$ to state $S_j$. Though the particular sequence which can be used for $T(S_i, S_j)$ is not important, still for convenience it is assumed that $T(S_i, S_j)$ always refers to the same sequence. Each of the sequences $T(Q_i, S_{i+1})$ requires at most $q - 1$ symbols, since the circuit has only $q$ states. Therefore, this portion of the checking sequence requires at most $q(q - 1) + (q + 1)L_{ds}$ symbols.

In the second part of the checking sequence, the transitions are checked in the order they appear in the state table. At the end of the first part of the checking sequence, the circuit will be in state $Q_1$, provided, in the first part it has produced the desired response. Thus the 0-transition out o state $S_1$ is checked by beginning the second part of the checking sequence with

$$T(Q_1, S_q)X_0T(Q_q, S_1)0X_0$$

If the 0-transition out of state $S_1$ leads to the state $S_j$, the circuit will now be in state $Q_j$. Thus to check the 1-transition out of state $S_1$ requires the application of the following sequence

$$T(Q_j, S_q)X_0T(Q_q, S_1)1X_0$$

In general, if the circuit is in a state $Q_i$ and it is desired to check the $X$-transition out of state $S_j$, the appropriate sequence will be

$$T(Q_i, S_{j-1})X_0T(Q_{j-1}, S_j)XX_0$$

The length of such a sequence need not exceed $2(q - 1) + 2L_{ds} + 1$. Thus all the transitions may be checked by a sequence whose length does not exceed $pq(2L_{ds} + 2q - 1)$, where $p$ denotes the number of input symbols of the machine.

Therefore, the total length of the checking sequence, according to Hennie, for a sequential circuit with a distinguishing sequence of length $L_{ds}$ need not exceed

$$pq(2L_{ds} + 2q - 1) + q(q - 1) + (q + 1)L_{ds} < 2q(p + 1)(L_{ds} + q)$$

Of course, this figure does not include the length of any initial synchronizing or homing sequence, which might add an additional $q(q + 1)/2$ symbols. Though the length of the distinguishing sequence $L_{ds}$ can, in the worst case, be an exponential function of $q$, most distinguishing sequences are of the order of $q$ or $q^2$; hence it becomes evident that the length of the checking sequence as constructed by Hennie, for the particular case when the machine possesses at least one distinguishing sequence, is found to be proportional to $q^2$ or $q^3$.

In all practical cases, instead of using this general synthesis procedure for designing checking sequences for a sequential machine having distinguishing sequences, many short cuts are usually taken into consideration. First, in many cases, the order in which the transitions are checked eliminates the necessity of the many $T$-sequences used in the general procedure. Secondly, once the design has reached a certain stage, some transitions will appear that have already been checked by the previous parts of the experiment, and thus will need no further checking of the same.

Let us illustrate the concepts by considering a simple example of a sequential machine $M_2$ as given in Table 3.3, which possesses a distinguishing sequence of 10.

Table 3.3 Machine $M_2$

| Present states | Next states and present outputs | |
|---|---|---|
| | $I_1 = 0$ | $I_2 = 1$ |
| $S_1$ | $S_3$, 0 | $S_2$, 0 |
| $S_2$ | $S_1$, 0 | $S_2$, 1 |
| $S_3$ | $S_2$, 1 | $S_3$, 0 |

The response of the different states of $M_2$ to the distinguishing sequence 10 is shown below in Table 3.3′.

The designing of the checking experiment starts as follows, where the first step sets the correctly operating circuit in the state $S_1$.

*Step 1*: Apply the input sequence 10, and observe the corresponding output response. If the response is 11, stop; the machine is faulty. If the response is 00, go directly to Step 2. If the response is 10, also go directly to Step 2. If the response is 01, apply a single input symbol 0, and then go to Step 2.

Table 3.3′ The response of the different states of $M_2$ to the distinguishing sequence 10 and the corresponding final states

| Initial state | Output sequence | Final state |
|---|---|---|
| $S_1$ | 00 | $S_1$ |
| $S_2$ | 10 | $S_1$ |
| $S_3$ | 01 | $S_2$ |

*Step 2*: This step deals with the formation of the checking sequence $X$ and simultaneous observation of the output $Z$ produced in response to it. The details of the procedure of formation of $X$ will be discussed shortly, but note that if the response to $X$ is not $Z$, then the machine can definitely be called faulty. But in case the response is $Z$, the obvious conclusion can be: either the machine $M_2$ is operating correctly, or else it has more than three states.

The above-mentioned Step 2, or the second part of the checking experiment begins with the basic assumption that if the machine $M_2$ is operating correctly, it must be in state $S_1$, after the first part of the experiment has been concluded. The input 10 is now applied to $M_2$ to get the response of state $S_1$ to the distinguishing sequence. If the machine operates correctly, it will again be in state $S_1$. In order to show the response of the state $S_2$ to the distinguishing sequence, it is necessary to apply 1 first, followed by 10. Next to obtain the response of $S_3$, the sequence 010 is applied. Of course, at this moment a sequence 0101 will be applied, the purpose of which will become soon clear. This gives also the response of $S_1$ for the second time to the distinguishing sequence, and leaves the machine in the state $S_2$ eventually. Up to this point, the checking sequence thus designed displays three different responses to the input sequence 10, and also confirms that if the machine operates correctly the input sequence 101 takes the machine from $S_1$ to $S_2$, while the sequence 100 takes it from $S_2$ to $S_3$ and then from $S_3$ to $S_1$. This is shown below.

| | | | | | | | | | | | | |
|---|---|---|---|---|---|---|---|---|---|---|---|---|
| $X$: | 1 | 0 | 1 | 1 | 0 | 0 | 1 | 0 | 0 | 1 | 0 | 1 |
| | $S_1$ | | | $S_2$ | | | $S_3$ | | | $S_1$ | | $S_2$ |
| $Z$: | 0 | 0 | 0 | 1 | 0 | 0 | 0 | 1 | 0 | 0 | 0 | 0 |

The second part of the checking sequence $X$ is now designed to verify whether the state transitions of the actual machine follow the state table behavior of the original nonfaulty machine. It begins with the checking of the 0-transition out of state $S_2$ in which the machine is now set after the application of the first part of the checking sequence $X$, and is accomplished by applying a 0 followed by 10. If the machine operates correctly, it should be in state $S_1$ at this moment. However, the experiment as constructed so far does not provide enough information to ensure that the machine is actually in state $S_1$ at this point. Only it can be stated that prior to the application of 10, it happened to be in state $S_1$. But by applying a single 1 at this moment it can be ascertained perfectly that the machine comes to the state $S_2$, since it is already known that the input sequence 101 takes the machine from $S_1$ to $S_2$. The 1-transition out of state $S_2$ can now be checked by applying another 1 and then following it with 10. Thus, in a similar manner, the remaining part of the checking sequence can be readily designed, and the entire transition-checking part of the experiment can be constructed as shown below.

$X$: 0 1 0 1 1 1 0 0 0 1 0 0 1 1 0 0 0 1 0 0 1 1 0

$S_2\ S_1$ $S_2\ S_2$ $S_3\ S_2$ $S_3\ S_3$ $S_1\ S_3$ $S_1\ S_2$

$Z$: 0 0 0 0 1 1 0 0 1 1 0 0 0 0 1 0 0 0 1 0 0 1 0

The length of the designed checking sequence $X$ is 35. It can be recalled here that the upper bound as obtainable using the general synthesis procedure is equal to 68, for $q = 3$, $p = 2$, and $L_{ds} = 2$.

Hennie's treatment for the design of checking experiment does not confine itself only to the discussion of machines having distinguishing sequences. As a follow up of the discussion, Hennie also developed techniques for the construction of checking experiments for those sequential machines that do not possess distinguishing sequences, but have characterizing sequences instead. For the machine $M_3$ as shown in Table 3.4, there is no distinguishing sequence, but the machine does have

Table 3.4 Machine $M_3$

| Present states | Next states and present outputs | |
|---|---|---|
| | $I_1 = 0$ | $I_2 = 1$ |
| $S_1$ | $S_2$, 0 | $S_4$, 0 |
| $S_2$ | $S_1$, 0 | $S_2$, 0 |
| $S_3$ | $S_4$, 1 | $S_1$, 0 |
| $S_4$ | $S_4$, 1 | $S_3$, 0 |

a characterizing set consisting of two sequences $X_1 = 0$, and $X_2 = 10$. The formation of the locating sequences for such machines was discussed earlier in Chapter 2. Recall that if we start at the state $S_1$, the sequence $[X_1T(Q_1, S_1)]^{q+1}X_2$ will be called a locating sequence for $S_1$ with respect to the characterising sequences $X_1$ and $X_2$. Similarly, the locating sequences for the other states can also be formed. In Table 3.5, the locating sequences of the machine $M_3$ for different states are listed. In actual practice, however, to serve our purpose, these locating sequences can be shortened, and the modified sequences are shown in the third column of the table.

This modification becomes obvious from the following. The checking sequence must contain one input 0 which causes the output 0, and if the actual circuit displays the same behaviour, it can have at most three states that can respond to 0 producing a 1. Thus the locating sequences for $S_3$ and $S_4$ should only contain four repetitions of 01 and 0, respectively. Also, the locating sequences for $S_1$ and $S_2$ need contain only three repetitions of 00, because the locating sequences for $S_3$ and $S_4$ show that there are two distinct states that produce 1 in response to 0 only.

Table 3.5 Locating sequences for different states of Machine $M_3$ with their corresponding output sequences

| States | Locating sequences | Modified locating sequences |
|---|---|---|
| $S_1$ | $L_1$: 00 00 00 00 00 10<br>00 00 00 00 00 01 | 00 00 00 10<br>00 00 00 01 |
| $S_2$ | $L_2$: 00 00 00 00 00 10<br>00 00 00 00 00 00 | 00 00 00 10<br>00 00 00 00 |
| $S_3$ | $L_3$: 01 01 01 01 01 10<br>10 10 10 10 10 00 | 01 01 01 01 10<br>10 10 10 10 00 |
| $S_4$ | $L_4$: 00 00 01 0<br>11 11 10 1 | 00 00 10<br>11 11 01 |

The appearance of a given locating sequence with its scheduled correct response at several places in an experiment guarantees that the machine is in the same state at the end of each appearance. To specifically determine in which state the machine places itself at the end of a locating sequence, it is required that the machine should be taken to that state twice; the state in question can then be identified by applying each of the characterizing sequences in turn. For this purpose, the following sequence is applied to the circuit at the state $S_1$, namely, $L_1$ 0 1 1 $L_1$ 10. Once the state of the machine at the end of a locating sequence becomes known, the transition resulting from that state by an input symbol can also be checked. As an example, the state $S_4$ in which the machine sets itself at the end of $L_1$ can be determined from the above-mentioned sequence by virtue of its responses to the characterizing sequences 0 and 10 at two different points in the experiment. The 0-transition out of $S_4$ can be checked now by applying the input sequence $L_1$ 00 11 $L_1$ 010. Here, the 0 following the first $L_1$ takes the machine from the state $S_4$ to an unknown state, say $S_X$. The next 0 measures the response of $S_X$ to one of the characterizing sequences. At this point, if the machine operates correctly, it will be again in state $S_4$, and it can be taken back to the state $S_1$ by the application of two 1's in succession. Thus, at this stage the locating sequence $L_1$ could be applied again, and the machine will come back to the state $S_4$. The 0-transition out of the state $S_4$ can also be made again, and following it, the second characterizing sequence can be applied. Therefore knowing the response for the two characterizing sequences, the state $S_X$ can definitely be identified.

In this order, the entire checking sequence can be constructed for a particular machine. For the machine of Table 3.4, the following sequence can well serve as the input of a complete checking experiment:

$$0\,1\,0\,1\,0\,1\,L_3\,L_2\,L_1\,L_4\,0\,L_4\,1\,0\,L_4\,0\,0\,L_4\,0\,1\,0\,L_4\,1\,1\,0\,0\,1\,L_4$$
$$1\,0\,0\,L_4\,1\,0\,1\,0\,L_4\,1\,1\,1\,0\,L_4\,1\,1\,0\,1\,0\,1\,L_4\,1\,1\,1\,1\,0\,1\,1\,0\,0\,0$$
$$0\,1\,L_4\,1\,1\,0\,0\,1\,0\,1\,1\,0\,1\,1\,0$$

Here, the sequence 010101 synchronizes the machine in state $S_3$, and all the four locating sequences can be applied in concatenation*, avoiding the use of transitional sequences in between. In the remaining part of the experiment, the use of $L_4$ is made justified on the ground that it is the shortest locating sequence, and hence its applications also help to bring down the total length of the experiment.

Hennie has also shown that for a $q$-state machine with $p$-input symbols where the characterizing sequences are made use of for the design of a checking experiment, the length of the checking experiment need not be greater than $pq^4(q + 1)!$. Further, he also discussed the problem of designing checking experiments for sequential machines taking into consideration the case where faults occurring in a machine may increase the number of states of the machine. For this purpose, he assumed that (i) the correctly operating machine is *strongly connected*; (ii) the number of states in the correctly operating circuit is less than twice the number of states in the reduced form of the correctly operating circuits and (iii) no malfunction increases the number of states to as many as twice the number in the reduced form. In the design of checking sequences under such criteria, the locating sequences are formed by repeating a given portion $q + i - 1$ times, if it would have ordinarily been repeated $i$ times only. Also, since the actual circuit will have more states than in the reduced form of the correctly operating circuit, it will not, in general, be possible to guarantee that the actual circuit has visited all its states. Hennie sought the solution for it by inserting a sequence of the form $U_{ij}$ in between two locating sequences $L_i$ and $L_j$, where $U_{ij}$ denotes a sequence that leads the circuit from the final state of $L_i$ to the initial state of $L_j$. Further, Hennie also showed that a checking sequence can be formed in this type of complicated cases where a fault will cause an increase in the number of states of a machine, by repeating the compound locating sequence $p^{3q-2}$, and in each time following it with a different sequence of length $3q - 2$.

Since the pioneering work of Hennie, the problem of fault detection and that of designing checking sequences for a sequential machines have been studied following two distinctly different kinds of approaches. In one

*If $\bar{I}_1$ and $\bar{I}_2$ be the two input sequences, then the concatenation $\bar{I}_1\bar{I}_2$ represents the sequence $\bar{I}_1$ followed by the sequence $\bar{I}_2$.

approach no consideration was given to modify a given sequential machine through augmentation of extra input or extra output terminals, and thereby adding some additional logic in the implementation. Kime [128], Kohavi and Kohavi [134], Gönenc [96], Hsieh [120], Kella [127], Farmer [73], Friedman and Menon [79], Boute [32], Das and Farmer [54], Das and Sheng [55], Sheng and Das [205, 206], and many others developed machine identification approaches that do not consider the use of any additional input or output symbols. On the other hand, Kohavi and Lavallee [133], Murakami et al. [160] and others initiated the approach based on augmenting a sequential machine by adding some extra input or output symbols, and showed that the problem of fault detection and diagnosis of a sequential machine can be made easier thereby. Later Holborow [116], Fujiwara and Kinoshita [84], Fujiwara et al. [85], and Sheppard and Vranesic [207] and others extended the idea of Kohavi and Lavallee and that of Murakami et al. The details of these works will be discussed in Chapter 4.

It must be borne in mind that methods for deriving test sequences for sequential machines have also been developed by other workers who did not restrict themselves to terminal measurements alone. In the works of Poage and McCluskey [169], Seshu and Freeman [101], and Roth et al. [186], it has been assumed that the circuit structure as well as the class of faults that can occur is known, though the advances in the technology of circuit modules and integrated circuits have limited these days, the experimenter's freedom to these types of measurements. In this class of measurements, it is customary to consider only single faults of the *stuck* types. It is also frequently assumed that the normal and the faulty circuits can be reset to a known initial state. The elimination of this assumption leads to additional computational complexity in the test derivation procedure and also results in longer test sequences.

### *3.3 Kime's method

Charles R. Kime [128] in one of his celebrated papers considered the problem of designing a checking sequence for a reducçd, *strongly connected* sequential machine which possesses a distinguishing sequence. His method was a modification of and a substantial improvement on Hennie's method and he obtained a reduced upper bound on the length of the checking sequence as compared to that obtained by Hennie. For a $q$-state sequential machine with $p$ input symbols, Kime's bound consists of $\left[(2L_{ds} + q - 1)(pq + 1) + \dfrac{pq(q + 1)}{2}\right]$ symbols, where $L_{ds}$ is the length of the distinguishing sequence. It can be recalled that for a similar situation, Hennie's bound was comprised of $[pq(2L_{ds} + 2q - 1) + q(q - 1) + (q + 1)L_{ds}]$ symbols. In general, the bound as obtained by Kime through

*Illustrations (equations etc.) cited in Section 3.3 are from the paper referred [128] in the bibliography and are presented with the permission of IEEE (copyright © 1966 IEEE).

his method, called a 'tree organized checking experiment', is smaller in length by

$$\left[(q-1)L_{ds} + \frac{pq(q-1)}{2} + q(q-2) + 1\right] \text{ symbols}$$

from that of Hennie. For $q \geqslant 2$, $p \geqslant 2$ and $L_{ds} \geqslant 1$, this difference is greater than or equal to four, and increases with $q$, $p$ and $L_{ds}$.

In the method of Kime, for a $q$-state, $p$-input machine having a distinguishing sequence $X_0$, the general form of the input sequence for the checking experiment consists of $pq + 1$ subsequences. The first subsequence is generally applied to the machine in the initial state, say $S_k$, and has the form $X_0X_0T(Q_l, S_k)$, where the state $Q_k$ results from the application of $X_0$ to $S_k$, and the state $Q_l$ results from the application of $X_0$ to $Q_k$. Any $T(S_i, S_j)$ denotes the minimal length sequence (input symbols) which takes the machine from the state $S_i$ to $S_j$. Here $T(S_i, S_j)$ always refers to the same sequence, once it has been selected. Noting for the moment that $S_{ij}$ is the state resulting from the application of input combination $I_i$ to the machine in state $S_j$, the remaining $pq$ subsequences can be obtained in one of the following two forms:

(i) if $S_{ij} \neq S_k$, then the subsequence is $X_0T(Q_k, S_j)I_iX_0T(Q_{ij}, S_k)$, and
(ii) if $S_{ij} = S_k$, then the subsequence is $X_0T(Q_k, S_j)I_i$.

Kime showed that for every state-input pair it is sufficient to include a subsequence either of the form (i) or of the form (ii) for verifying the output, and also the next state transition. These $pq + 1$ subsequences form a tree-organized checking sequence, where any transitional subsequence $T(Q_k, S_j)$ can be read from the tree, because the $pq + 1$ subsequences contain in them two basic requirements for a tree-organized structure to be built, namely, (a) for each state-input pair $(S_j, I_i)$ there should exist a subsequence $X_0X'I_iX_0$, such that if $X_0X'I_i$ is applied to the machine in state $S_k$, the resultant state becomes $S_{ij}$, (b) for each state $S_j$, there exists a subsequence $X_0X'X_0$ such that $X'$ is the same sequence as for the corresponding state $S_j$ in (a), and thus if $X_0X'$ is applied to the machine in state $S_k$, the resulting state becomes $S_j$.

The logic for inclusion of the condition (a) as stated above in the subsequences is obvious from the description of the checking experiment. However, to justify that the $pq + 1$ subsequences also include the condition (b) requires some clarification. For the state $Q_k$, $X' = T(Q_k, Q_k) = \lambda$, the null sequence. Thus $X_0X'X_0 = X_0X_0$ is included in the first subsequence. Next, for the set of states $\{S_{t_1}, S_{t_2}, \ldots, S_{t_p}\}$ where $S_{t_i} = f(Q_k, I_i)$, $f$ being the next state function, consider $X_0X'X_0 = X_0T(Q_k, S_{t_i})X_0 = X_0I_iX_0$ a subsequence to check the transition of $I_i$ from state $Q_k$. Consider finally a state $S'$ such that $T(Q_k, S')$ is a transitional subsequence of minimal length $l$ which is used in the experiment. Also, take a set of states $\{S'_{t_1}, S'_{t_2}, \ldots, S'_{t_p}\}$ such that (i) $S'_{t_i} = f(S', I_i)$, and (ii) there exists no $T(Q_k, S_{t_i})$ of length less than or equal to $l$. Now if we take $X'$ for $S'_{t_i}$ as $T(Q_k, S')I_i$ for any $I_i$ satisfying the conditions, then $T(Q_k, S'_{t_i})$ will appear to be $T(Q_k, S')I_i$ which is of length $l+1$. Therefore, $X_0X'X_0$ for $S'_{t_i}$ is equal to $X_0T(Q_k, S')I_iX_0$ which is a subsequence used for checking the transition of $I_i$ from $S'$. The aforesaid argument follows obviously because the machine is strongly connected, and hence for each state $S_j$ there exists $X_0X'X_0$ which is $X_0X'I_iX_0$ for some other state $S'$.

## 3.4 Method based on variable length distinguishing sequences

The use of the properties of variable length distinguishing sequences (VLDS) for the construction of efficient fault detection experiments has been considered by I. Kohavi and Z. Kohavi [134]. The VLDS was discussed earlier briefly in Chapter 2. A VLDS possesses different shortest distinguishing prefixes (SDP) for the different states of the machine. An SDP for any state can equally serve the purpose of identification of that state in a checking experiment, and can also eliminate the use of distinguishing sequences. The usefulness of the VLDS becomes significant in the design of checking sequences whenever its average length is shorter than that of the fixed length distinguishing sequences (FLDS). In addition, the checking sequences become shorter if a large number of next state entries possess shorter prefixes of the VLDS. It is to be noted here that every sequential machine may not have a VLDS. But the possibility of having a VLDS for a sequential machine increases with the increase in the number of output symbols. These disadvantages have limited the scope of utilization of VLDS to the design of fault detection experiments in sequential machines.

## *3.5 Graph theoretic and algorithmic design of Gönenc

G. Gönenc [95] has presented a methodical approach for the organization of fault detection experiments of synchronous sequential machines possessing distinguishing sequences. Apart from the algorithmic nature of the procedure, it also gives an upper bound on the length of the experiment that is smaller than any of the bounds in the earlier works in this direction. For a $q$-state and $p$-input sequential machine, the upper bound of his method is

$$[(1+p)(q-1)^2 + 2qL_{ds} + pq(L_{ds}+1)] \text{ symbols,}$$

where $L_{ds}$ is, as before, the length of the distinguishing sequence.

In the method of Gönenc, a state is recognized in three different ways: (i) A state is said to be a $d$-recognized state $S_i$, if in response to the DS (distinguishing sequence) $X_d$, the resultant output $Z_i$ identifies uniquely the state $S_i$. These states are marked with a superscript $d$. (ii) Since every DS is also an HS (homing sequence), a state of the machine after the application of $X_d$ can be identified as $Q_i$. Such a state is called $q$-recog-

*Illustrations (equations, tables and figures) in the Section 3.5 are from the paper referred [95] in the bibliography and are presented with the permission of IEEE (copyright © 1970 IEEE).

nized and shown with a superscript $q$. (iii) If in the course of an experiment, in response to the input sequence $X_dXX_d$, an output sequence $Z_iZZ_j$ is observed where $X$ is any arbitrary sequence, then it can be said that before the application of $X$, the machine was in a $q$-recognized state $S_k = Q_i$, and after the application of $X$, it is in a $d$-recognized state $S_j$. It establishes a relationship $S_j = f(S_k, X)$, where $f$ stands for the next state function. Such a state $S_j$ is said to be a $t$-recognized state, and is shown with a superscript $t$.

The checking sequence, as organized in this method, consists of two parts known as $\alpha$-sequence part and $\beta$-sequence part. The $\alpha$-sequence part is organized basically to serve two purposes, namely, (i) to recognize all states, and (ii) to determine all $Q$'s. The $\beta$-sequence is organized to check each transition in the state table. This sequence consists of subsequences which are called cells. A cell to check the transition under the input $I_i$ belonging to the input alphabet is the input sequence $I_iX_d$ applied to the machine in state $S_j$. Since the end state of each cell is readily $q$-recognized from the knowledge about the $Q$'s as obtained from the $\alpha$-sequence, this end state may well be considered as the beginning state of the next cell, and thus amounts to the reduction of overall checking sequence. Let us take the machine $M_4$ as shown in Table 3.6 as an illustration.

Table 3.6 Machine $M_4$

| Present states | Next states and present outputs | |
|---|---|---|
| | $I_1 = 0$ | $I_2 = 1$ |
| $S_1$ | $S_2$, 0 | $S_6$, 0 |
| $S_2$ | $S_3$, 0 | $S_5$, 1 |
| $S_3$ | $S_4$, 1 | $S_2$, 0 |
| $S_4$ | $S_5$, 0 | $S_1$, 0 |
| $S_5$ | $S_6$, 0 | $S_4$, 1 |
| $S_6$ | $S_2$, 1 | $S_5$, 1 |

The designing of the $\alpha$-sequence for the machine in Table 3.6 goes as follows: We take the shortest distinguishing sequence 001 as $X_d$, and the $X_d$-table and the $X_d$-diagram are constructed as shown in Table 3.7 and in Fig. 3.1, respectively.

The $X_d$-diagram is the transition diagram obtained from the $X_d$-table, whereas an $X_d$-table gives the values of the next state function $f(S_i, X_d)$ and of the output function $g(S_i, X_d)$ for all $i$, $i = 1, 2, \ldots, q$, and the input string $X_d$. The states $S_3$ and $S_6$ are called here sources. (A source for an input string $X$ is a state which is not the next state of any state under input $X$.) The convention is to take any one of the sources as the

Table 3.7 $X_d$-table

| $S_i$ | $Q_i$ | $Z_i$ |
|---|---|---|
| $S_1$ | $S_2$ | 000 |
| $S_2$ | $S_1$ | 010 |
| $S_3$ | $S_4$ | 101 |
| $S_4$ | $S_5$ | 001 |
| $S_5$ | $S_5$ | 011 |
| $S_6$ | $S_2$ | 100 |

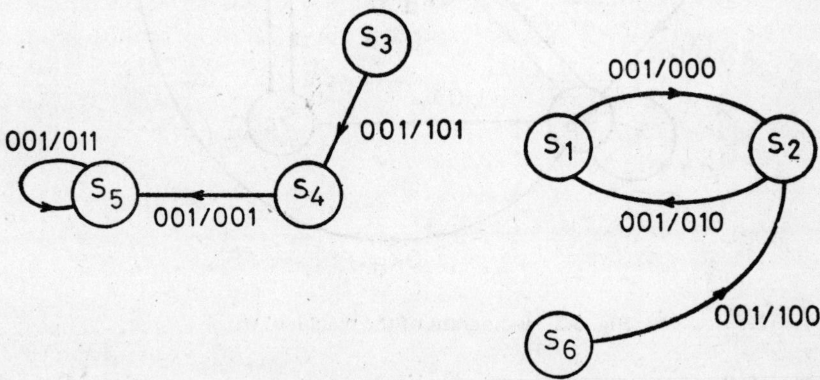


Fig. 3.1 $X_d$-diagram of the machine $M_4$.

starting state. The machine can be initially placed in state $S_3$ by the application of the synchronizing sequence 11011100. According to the algorithm as given in this context, if we form the $\alpha$-sequence starting from the state $S_3$, it will require a transfer sequence $T(5, 6) = 0$, because we have two sources in the $X_d$-diagram, and it is a disjoint graph of two subgraphs. The $\alpha$-sequence organized in this connection is shown below:

| Input: | | 001 | | 001 | | 001 | | 001 | | 0 | | 001 | | 001 | | 001 | | 001 | |
|---|---|---|---|---|---|---|---|---|---|---|---|---|---|---|---|---|---|---|---|
| State: | $S_3$ | | $S_4^d$ | | $S_5^d$ | | $S_5^d$ | | $S_5^q$ | | $S_6^d$ | | $S_2^d$ | | $S_1^d$ | | $S_2^d$ | | $S_1^q$ |
| Output: | | 101 | | 001 | | 011 | | 011 | | 0 | | 100 | | 010 | | 000 | | 010 | |

The $\beta$-sequence for the above machine is organized to check all the transitions, and it is composed of cells and $T$-sequences. Similar to the $X_d$-diagram as described earlier, a $\beta$-diagram is formed for this purpose. The $\beta$-diagram is obtained by considering inputs $I_iX_d$ for all $I_i \in I$, or, stated in other words, it is the $(\bigcup_{I_i \in I} I_iX_d)$-diagram. The $\beta$-diagram for the machine of Table 3.6 is shown in Fig. 3.2.

In a $\beta$-diagram the state alphabet $S$ can be divided into subsets $R$, $F$

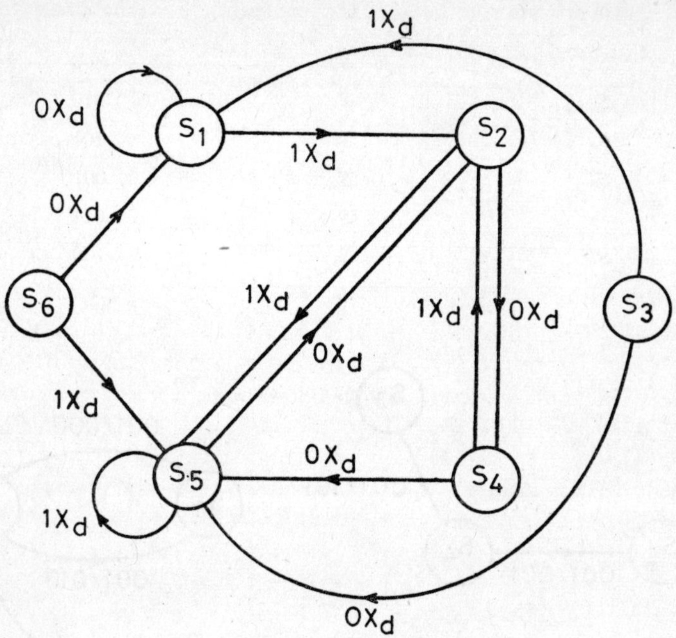


Fig. 3.2 $\beta$-diagram of the machine $M_4$.

and $P$ as follows:

$$R = \{S_i \in S \mid d_i^+ = d_i^-\},$$

$$F = \{S_i \in S \mid d_i^+ > d_i^-\},$$

$$P = \{S_i \in S \mid d_i^+ < d_i^-\},$$

where $d_i^+$ is the number of arcs leaving the state $S_i$, and $d_i^-$ is the number of arcs entering the state $S_i$. Evidently $d_i^+ = p$, for all $i$, where $p$ is the total number of input symbols in the input alphabet. A graph is said to be Eulerian if $d_i^+ = d_i^-$ for all $i$. Therefore, for an Eulerian graph the sets $F$ and $P$ are empty and $R = S$, the entire state alphabet of the machine. Now if additionally we observe that the set of arcs is a circuit (a finite path in which the initial vertex coincides with the terminal vertex) it then constitutes the unique minimal covering. Here by the term covering we mean partitioning of all the arcs of a graph into simple paths which are disjoint (i.e., not having any common arc). But for a directed graph which is connected but not Eulerian, every minimal covering of the graph consists of $K$ paths which joins a vertex in $F$ with a vertex in $P$, where

$$K = \sum_{S_i \in F} (d_i^+ - d_i^-) = \sum_{S_i \in P} (d_i^- - d_i^+)$$

Gönenc has discussed the problem of designing the $\beta$-sequence organization based on finding a minimal covering of the $\beta$-diagram. The algorithm given for the purpose is as follows:

*Step 1*: Start from any state in $F$.

*Step 2*: Each time an are is followed, it is to be erased next.

*Step 3*: In no time the use of that particular arc has to be made of which the deletion usually divides the graph into two components, provided there also exists any other arc which can be followed.

*Step 4*: When it is not possible to proceed further, again a new starting from a state in $F$ will be made so that from each state $S_j \in F$, exactly $\lambda_j$ starts will be made, where $\lambda_j = d_j^+ - d_j^-$.

It is evident from the above steps that for a restart, a 'jump', i.e., a $T$-sequence of the form $T(S_i, S_j)$ is used, where $S_i \in P$ and $S_j \in F$. For $K$ paths from $P$ to $F$, there always exist $(K - 1)$ $T$-sequences.

In the case of the machine of Table 3.6 as cited before, $F = (S_3, S_4, S_6)$, and $P = (S_1, S_2, S_5)$, and $K$ becomes equal to $\lambda_3 + \lambda_4 + \lambda_6 = 2 + 1 + 2 = 5$. This indicates that the minimal covering consists of five paths and four jumps if the $\beta$-sequence starts from the states $S_3$, $S_4$, or $S_6$. The same consists of six paths and five jumps if the $\beta$-sequence starts from any other states. In the $\beta$-sequence two $(\lambda_3)$ paths start from state $S_3$, one $(\lambda_4)$ path starts from state $S_4$ and two $(\lambda_6)$ paths start from state $S_6$. An additional path starts from the starting state of $\beta$, if the said starting states are $S_1$, $S_2$ or $S_5$.

Three of the possible minimal coverings are shown below.

(1) $S_4S_5S_5S_2S_5$-$S_6S_5$-$S_6S_1S_1S_2S_4S_2$-$S_3S_1$-$S_3S_5$

(2) $S_3S_1S_1S_2S_5S_5S_2S_4S_2$-$S_3S_5$-$S_4S_5$-$S_6S_5$-$S_6S_1$

(3) $S_1S_1S_2S_5S_5S_2S_4S_2$-$S_3S_5$-$S_4S_5$-$S_6S_5$-$S_6S_1$-$S_3S_1$

The state sequences are separated here by hyphens corresponding to jumps ($T$-sequences). In the last covering, 6 paths and 5 jumps are taken, because the starting state here falls in the set $P$, and not in $F$, and it needs one additional path as well as jump. For the construction of checking experiment it is desirable that the starting state of the $\beta$-sequence should be the end state of the $\alpha$-sequence. This indicates that the input string of the $\beta$-sequence corresponding to the last covering will be suitable for the checking experiment. The $\beta$-sequence corresponding to the last covering is shown below:

$$0X_d\ 1X_d\ 1X_d\ 1X_d\ 0X_d\ 0X_d\ 1X_d\ 00X_d\ 10X_d\ 01X_d\ 00X_d\ 001X_d.$$

## *3.6 Hsieh's method

A new approach in connection with the design of an efficient checking experiment has been proposed recently by E. P. Hsieh [120], who intro-

*Illustrations (equations etc.) cited in Section 3.6 are from the paper referred [120] in the bibliography and are presented with the permission of IEEE (copyright © 1971 IEEE).

duced the concept of four types of sequences, namely, the compound DS, the resolving sequence (RS), the compound RS, and the simple $I/O$ sequence as discussed earlier in Chapter 2. In this new technique, the checking experiment $T$ is considered functionally subdivided into three parts as: $T = T_I + T_d + T_v$, referred to as the initialization part, the diagnosing part, and the validating part, respectively. The initialization part of the experiment brings the machine $M_t$ under test to the correct starting state for the experiment, regardless of the initial state of $M_t$. In the validating part, a selected set of simple $I/O$ sequences of the correctly operating machine $M$ is found to be valid for the machine $M_t$ under test. It is needed to be sure in the validating part that at the instant when the fault will occur, the application of the simple $I/O$ sequence $\pi$ will still justify that the machine must have been in the same state after each occurrence of $\pi$. Finally, the diagnosing part of the experiment characterizes the machine under test based on the use of the set of simple $I/O$ sequences validated by $T_v$.

For machines having a DS or compound DS, the validating part $T_v$ can always be made completely contained in the diagnosing part $T_d$, and in such cases the validating part does not increase the overall length of the checking experiment. But for more general class of machines, the validating part usually results in an increase of the length of the checking experiment significantly. Under such circumstances, the set of $I/O$ sequences essential to $T_v$ and the set of sequences essential to $T_d$ can be shared and merged in an optimal way to form the entire checking experiment.

The concept of the property-validating experiment as a functional part of the checking experiment not only verifies that the machine under test possesses certain useful sequences known to be valid homing sequences, but also provides a method of utilizing these valid homing sequences for the construction of efficient checking experiments. A valid homing sequence of a machine $M$ is an $I/O$ sequence generated by $M$ whose occurrence alone in a terminal experiment ensures that the machine under test is in a unique final state. In this method, for a large class of machines without DS, a significant reduction in the length of the checking experiments is achieved without employing any extra terminal or augmented logic. For a $q$-state, $p$-input machine with compound DS, RS, and compound RS, the ratio of the improvement is approximately $q!$ over the best bound of $pq^4[(q+1)!]$ as established earlier by Hennie. For machines with simple $I/O$ sequences, the ratio of improvement is approximately $(1/12)p(4/e)^{t+1}t!$, with $t = [(q+1)/2]$. In the latter part of his paper, Hsieh [120] also extended the concept of validating experiments to include those faults which may cause an increase in the number of states of the machine. It was admitted in the paper that, in general, no finite preset terminal experiment can establish the equivalence of the terminal behaviour of a given machine $M^*$ with the correct behavior as described by a given reduced machine $M$, if a fault occurs such as to cause the faulty machine $M^*$ to have more than twice the number of states of $M$. If a fault increases the number of states of the machine to within the limit of twice the number of states of the original nonfaulty machine, then a bound of $p^{\Delta q+1}q \cdot B$ can be achieved, where the maximum anticipated increase of the number of states due to faults is $\Delta q$, with $B$ being the bound for checking experiments designed strictly under the assumption that no faults increase the number of machine states. The exponent on $p$ in the bound is independent of the number of states $q$ of the given state table, which represents a significant improvement over the bound given earlier by Hennie [113].

### 3.7 Method based on restricted checking sequences

It has been subsequently demonstrated by A. D. Friedman and P. R. Menon [79] that the imposition of some restrictions on the realization of a sequential machine facilitate the derivation of efficient checking sequences for certain classes of machines. It was assumed in their paper that the state logic or the output logic may be faulty, but both of them cannot be faulty at the same time. This is justifiable under the single fault assumption (as widely used in the circuit testing approach), if there is no shared logic between the next state and output circuits. It was also assumed in this method that the normal machine had a synchronizing sequence. The class of faults considered is more restricted than that in the machine identification approach, but more general than that in the circuit testing approach. These restrictions on the realization and on the class of faults lead to shorter checking experiments in many cases.

The circuit testing approach in the detection of faults in sequential machines does not seem to be of much practical importance with the recent developments in the technology of circuit modules and integrated circuits. In sequential machines, there can always be some shared logic between the next state and output transitions, and also all machines do not possess synchronizing sequences. Thus, the above idea does not seem to be that versatile though it is based following the approach of terminal measurements.

### *3.8 Identification approach developed by Farmer

D. E. Farmer [73] has studied the problem of fault detection and that of designing checking experiments for sequential machines with a particular emphasis on the type of machines which do not have distinguishing sequences. He obtained a shorter length of the checking experiment mainly through the use of the following ideas: (i) identifying each state with its own input/output set rather than using a common set for all states; (ii) utilizing overlapping of the required input/output sequences so that a portion of the experiment serves more than one purpose; and finally (iii) verifying the reference condition in which the machine is placed at many points in the experiment by using as short a locating sequence as possible.

*Illustrations (equations, tables and figures) in Section 3.8 are from the paper referred [73] in the bibliography and are presented with the permission of IEEE (copyright © 1973 IEEE).

Some important distinctions are made between the types of locating sequences introduced in the works of Hennie, and those defined and used by Farmer. A particular type of locating sequence which was used by Hennie is called as a characterization-verification sequence. A characterization-verification sequence for a state $S_i$ of the tested machine $M'$ is an input sequence for which the observation of a specified response verifies that $M'$ possesses a state characterized by the characterizing set for $S_i$ of the fault-free machine $M$. On the other hand, the locating sequence as used by Farmer is an input sequence for which the observation of a specified response permits the determination of the state of the machine at some known points in the sequence. Though Farmer has used several such types of sequences in the experiment, yet each serves a different purpose. An initial state locating sequence for a state $S_i$ is obtained by selecting the input sequence corresponding to a path traced from the zero level node to a node labelled with an initial state uncertainty partition in which $S_i$ occurs as a singleton block. On the other hand, a present state locating sequence for a state $S_i$ is obtained by selecting the input sequence corresponding to a path from the zero level node to a node labelled with a present state uncertainty vector in which $S_i$ occurs as a singleton or as a homogeneous component.

The following notation is introduced for the purpose of simplifying the characterization-verification sequence.

Suppose $\{X_{i1}, X_{i2}, \ldots, X_{il}, \ldots, X_{ik}\}$ denote a characterizing set for the state $S_i$ of the machine $M$, $1 \leqslant l \leqslant k$, and let $\{Q_{i1}, Q_{i2}, \ldots, Q_{il}, \ldots, Q_{ik}\}$ denote the set of states in which $M$ is left following the application of $X_{il}$, $M$ being initially in the state $S_i$.

Let $T(S_a, S_b)$ denote the input sequence required to transfer the machine from state $S_a$ to state $S_b$, and let $\overline{X}_{il} = X_{il}T((Q_{il}, S_i)$. Then $\{\overline{X}_{i1}, \overline{X}_{i2}, \ldots, \overline{X}_{il}, \ldots, \overline{X}_{ik}\}$ will also be called a characterizing set for the state $S_i$, if it is applied to $M$ initially in $S_i$. The set of responses of $M$ to the element $X_{il}$ of the modified characterizing set for $S_i$ is denoted by $\{\overline{Z}_{i1}, \overline{Z}_{i2}, \ldots, \overline{Z}_{il}, \ldots, \overline{Z}_{ik}\}$. An expression has been found for the characterization-verification sequence for state $S_i'$ of the tested machine $M'$ having the total number of states $q$ as follows:

Let

$$Y_{i1} = \overline{X}_{i1}$$

$$Y_{i2} = Y_{i1}^q \overline{X}_{i2}$$

$$Y_{i3} = Y_{i2}^q Y_{i1}^q \overline{X}_{i3}$$

$$\vdots$$

$$Y_{il} = Y_{i(i-i)}^q Y_{i(l-2)}^q \ldots Y_{il}^q \overline{X}_{il}$$

$$\vdots$$

$$Y_{i(k-1)} = Y_{i(k-2)}^q Y_{i(k-3)}^q \ldots Y_{il}^q \overline{X}_{i(k-1)}$$

$$Y_{ik} = Y_{i(k-1)}^q Y_{i(k-2)}^q \ldots Y_{il}^q \overline{X}_{ik}$$

and this expression of $Y_{ik}$ as obtained above is a characterization-verification sequence for state $S'_i$ of the machine $M'$.

It is the purpose of a fault detection experiment to establish if the behavior of the tested machine $M'$, initially in state $S'_i$, is or is not distinguishable from the behavior of the fault-free machine $M$ initially in $S_i$, for all $i$. The experiment is divided in two parts: (i) The characterizing portion which establishes that the tested machine has a state set that is characterized by the characterizing sets for the state set of the fault-free machine. It also selects a state, readily locatable, for the purpose of utilizing it as a base state for the second part of the experiment. (ii) The transition checking portion is designed and included in the second part in the usual manner.

The designing of the first part of the experiment differs depending upon whether the machine possesses an initial state locating sequence or not. If an initial state locating sequence exists, it is generally included as an element of the characterizing set for that state so that its verification can be made simultaneously with that for the characterizing sets of the various states. The verification of all the characterizing sets by the application of characterization-verification sequence for each state then proceeds similar to that as in the case of Hennie's method.

Now for the case of machines which do not have an initial state locating sequence, the organization of the characterizing portion verifies the presence of a shortest present state locating sequence $X$ (with its corresponding output $Z$) for a state $S_i$ of the machine. This can be verified readily, if it can be shown that the states of the tested machine $M'$ subdivide themselves into three categories as follows:

*Subset 1*: Those states for which the machine has a transition under input $X$ into $S_i$ accompanied with the response $Z$.

*Subset 2*: Those states for which the machine has a transition under input $X$ into $S_i$, but not accompanied with the response $Z$.

*Subset 3*: Those states for which the machine has a transition under input $X$ into some state other than $S_i$, but not accompanied with the response $Z$.

To show that $X$ is a present state locating sequence for state $S'_i$ of the tested machine $M'$, it is required to prove that the states of the tested machine fall under the above three stated subsets. This is done as follows:

*Step 1*: Each characterization-verification sequence returns $M'$ to the respective initial state which is being ascertained here. The idea is to include in the characterizing portion a number of characterization-verification sequences for each state equal to the order of the characterizing set for that state, with each such sequence suffixed with a different characterizing set element.

*Step 2*: For each state $S'_j$ belonging to Subset 1, and for each element of the characterizing set for $S'_j$, we need to include a characterization-verification sequence for $S'_j$ (verified by Step 1) to return $M'$ to $S'_j$ followed by $X$, and in turn followed by the element of the characterizing set for $S'_i$.

*Step 3*: For each state belonging to Subsets 2 and 3, we have to include also in the characterizing portion a characterization-verification sequence followed by $X$.

At this point, it will be appropriate to illustrate how this method can be applied in the case of an actual machine as shown in Table 3.8.

Table 3.8 State table and characterising set for machine $M_5$

| Present states | Next states and present outputs | | Characterizing set | Response to characterizing set |
|---|---|---|---|---|
| | $I_1 = 0$ | $I_2 = 1$ | | |
| $S_1$ | $S_1$, 1 | $S_3$, 0 | {0, 10} | {1, 00} |
| $S_2$ | $S_2$, 0 | $S_3$, 0 | {0, 10} | {0, 00} |
| $S_3$ | $S_2$, 0 | $S_4$, 0 | {0, 10} | {0, 01} |
| $S_4$ | $S_1$, 1 | $S_4$, 0 | {0, 10} | {1, 01} |

The above machine has no initial state locating sequence. The sequence 0 is a present state locating sequence for the state $S_1$ as long as it gives an output of 1. For this machine, Subset 1 consists of states $S_1$ and $S_4$, Subset 2 is empty, and Subset 3 consists of states $S_2$, $S_3$. The characterization-verification sequences and their associated responses for the states of the machine are shown in Table 3.9.

Table 3.9 Characterization-verification sequences for machine $M_5$

| States ($S_i$) | Characterization-verification sequences ($Y_i$) | Responses of the machine ($W_i$) |
|---|---|---|
| $S_1$ | $Y_1 = (10110)^4 0$ | $W_1 = (00001)^4 1$ |
| $S_2$ | $Y_2 = (10)^4 0$ | $W_2 = (00)^4 0$ |
| $S_3$ | $Y_3 = (101)^4 0$ | $W_3 = (010)^4 0$ |
| $S_4$ | $Y_4 = (1011)^4 0$ | $W_4 = (0100)^4 1$ |

With these tools in hand, the characterizing portion of an experiment can be designed to verify the present state locating sequence 0 as shown below. Some advantage is taken of overlappings in Steps (1) and (2), or Steps (1) and (3), and have been indicated by placing the step numbers with the segments to which they apply.

Input : $\overbrace{Y_1\ \ 0}^{(1)}\ \ 0\ \ \overbrace{Y_1\ \ 10}^{(1)}\ \ 110\ \ Y_1\ \ 0\ \ 10$

Fault-free state : $S_1\ \ S_1\ \ S_1\ \ S_1\ \ S_1\ \ S_2\ \ S_1\ \ S_1\ \ S_1$ (Contd.)

Output : $\underbrace{W_1\ \ 1\ \ 1}_{(2)}\ \ W_1\ \ 00\ \ 001\ \ \underbrace{W_1\ \ 1\ \ 00}_{(2)}$

Input : $\overbrace{Y_2\ \ 0}^{(1)}\ \ \overbrace{Y_2\ \ 10}^{(1)}\ \ 1\ \ \overbrace{Y_3\ \ 0}^{(1)}\ \ 1\ \ \overbrace{Y_3\ \ 10}^{(1)}$

Fault-free state : $S_2\ \ S_2\ \ S_2\ \ S_2\ \ S_2\ \ S_3\ \ S_3\ \ S_2\ \ S_3\ \ S_3$ (Contd.)

Output : $\underbrace{W_2\ \ 0}_{(3)}\ \ W_2\ \ 00\ \ 0\ \ \underbrace{W_3\ \ 0}_{(3)}\ \ 0\ \ W_3\ \ 01$

Input : $11\ \ \overbrace{Y_4\ \ 0}^{(1)}\ \ 0\ \ 11\ \ \overbrace{Y_4\ \ 10}^{(1)}\ \ 11\ \ Y_4\ \ 0\ \ 10$

Fault-free state : $S_1\ \ S_4\ \ S_4\ \ S_1\ \ S_1\ \ S_4\ \ S_4\ \ S_1\ \ S_4\ \ S_4\ \ S_1\ \ S_2$

Output : $00\ \ \underbrace{W_4\ \ 1\ \ 1}_{(2)}\ \ 00\ \ W_4\ \ 01\ \ 00\ \ \underbrace{W_4\ \ 1\ \ 00}_{(2)}$

As a final part of the fault-detection experiment, the concept of a transition-checking tree was introduced for the construction of the transition-checking portion of the experiment. This tree is in the form of state transition diagram with nodes representing states, and directed branches representing transitions for specified input sequences. It consists of four segments as follows:

(i) The *locating sequence root* consists of a single branch originating at the base state $S_0$ and directed to state $Q_0$. The branch is labelled with the input sequence $X_0$.

(ii) The *transition covering portion* consists of a state transition diagram originating at $Q_0$ and developed in the form of a tree manner through successive levels, until all the transitions of the machine have been covered.

(iii) The *characterization-completion portion* carries the tree through an additional level. Each terminal node of the transition covering portion is augmented here with a branch or branches labelled with input sequence or sequences required to provide that each node of the transition-covering portion be succeeded by tree paths corresponding to each element of the characterizing set for the state associated with that node.

(iv) The *closing portion* of the tree carries the tree through an additional level so that each tree path ends at the base state $S_0$.

In the case of machines that do not have initial state locating sequences for any of their states, it is necessary to use a present state locating sequence for the construction of the transition-checking tree. In this case, the tree originates at a node corresponding to the subset of states for which the machine reaches $Q_0$ on application of $X_0$, the present state locating sequence. The closing portion of the tree returns the machine to the most convenient state of this subset. As an example, the transition-checking tree for the machine $M_5$ as shown in Table 3.8 is obtained as

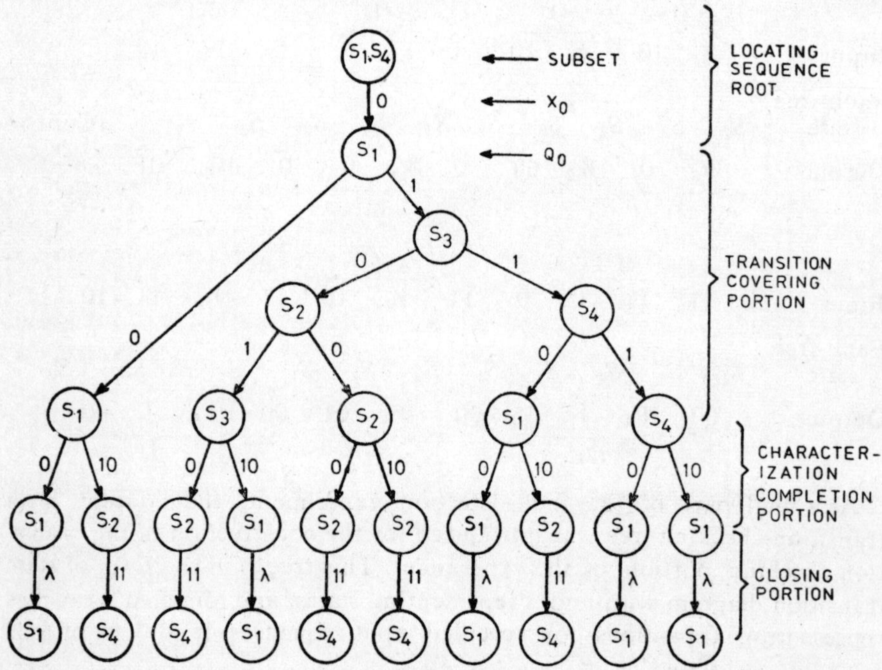


Fig. 3.3 Transition-checking tree for machine $M_5$.

follows. The transition-checking portion of a fault-detection experiment can be obtained very easily from such a tree. This is usually done by tracing all the paths through the tree from the origin $S_0$ back to the common destination, also $S_0$. It becomes at once obvious that for the tree shown as here, the input sequence for the transition checking experiment consists of the following ten subsequences: (000) (001011) (0101011) (010110) (0100011) (01001011) (01100) (01101011) (01110) (011110).

## 3.9 Method of Sheng and Das

In one of the most interesting approaches developed by C. L. Sheng and S. R. Das [206], the problem of identification of a reduced, completely specified, strongly connected synchronous sequential machine was studied under the constraint that for a certain given input sequence applied to the machine, the corresponding output sequence is known, while the flow table of the machine is not given. The study basically involves transformation of the problem of identification to one of reducing the flow table of an incompletely specified, *nonstrongly* connected machine by merging of states. Although the flow table of the machine is not provided, it was shown that the flow table of the corresponding incompletely specified machine, called a *maximum flow table* (a discussion on maximum flow table is included in Chapter 2), can be readily constructed from the given input sequence and its corresponding output sequence, from which the reduced flow table of the machine can eventually be obtained.

The problem of constructing the input/output sequences that will be sufficient to identify a given machine $M$, with a known flow table $F$, which is identical to the problem of designing checking experiments for sequential machines was also considered by the same authors. The authors also demonstrated earlier [55, 205] that a necessary condition for an input sequence $X_i$ to be able to identify a given sequential machine $M$ is that $X_i$ must lead $M$ through all its state transitions. The first segment of the input/output sequences that will be sufficient to identify the given machine $M$ is constructed following the aforesaid criterion of leading the machine through all its state transitions. The length of this first segment $X_{ia}$ is either greater than or equal to $pq$, where $p$ and $q$ denote the numbers of input symbols and states of the machine, respectively. Some information, if not all, regarding the machine can thus be obtained from this first segment, and this information helps to choose the next segment of the input/output sequences. In this way, with more and more information obtained from the various segments of the input/output sequences, the machine is eventually identified by the combined input/output sequences. The method is thus obviously an adaptive procedure.

The selection of the second and subsequent segments of the input/output sequences can be done quite readily. With the first segment of the input/output sequences obtained, a maximum flow table $F_1'$ is formed. This table, on reduction, yields, in general, more than one completely specified reduced flow tables $\{F_i\}$, one of which is $F$ of $M$. After these reduced flow tables $\{F_i\}$ are obtained, from an inspection of these tables, an input sequence of minimal length is selected such that the output sequence from $F$ will be different from that of at least one other table in $\{F_i\}$, to discriminate $F$ from those of $\{F_i\}$ that give different outputs. This sequence is successively augmented to make it equal to $X_{ib}$ to discriminate $F$, if possible, from all the remaining flow tables of $\{F_i\}$. If all of $\{F_i\}$ are completely specified, then $F$ will be discriminated from all the remaining tables of $\{F_i\}$ by $X_{ib}$; if not, $F$ may still be discriminated with $X_{ia}X_{ib}$ and its corresponding output sequence representing in each case the identifying sequences of $M$. If all of $\{F_i\}$ are not completely specified and $F$ cannot be discriminated from all the remaining tables of $\{F_i\}$ by $X_{ib}$, then the same procedure is repeated by constructing another maximum flow table $F_2'$ with respect to $X_{ia}X_{ib}$ and its corresponding output sequence, until finally the checking sequence for $M$ is obtained.

### 3.10 Boute's concept of optimum state-identification

A new concept, that of a distinguishing set or *D*-set, was introduced by R. T. Boute [32], and it was shown that the sequences of a distinguishing set actually constitute the optimum length sequences necessary for state identification through input/output observations only. The definition of a *D*-set has been provided in Chapter 2. Utilizing this concept of *D*-set, the checking sequence for a machine can be made quite short in length. This method includes an implicit transition-verification approach with increased possibility of 'telescoping'. The word 'telescoping' implies that if two sequences are to be applied during an experiment, and the final part (suffix) of one sequence is the same as the initial part (prefix) of the other sequence, then the experiment can be organized in such a way that this common part has to be applied only once. The main advantage of this approach is that if in some machines there does not exist any preset distinguishing sequence, but there exists a distinguishing set, then this *D*-set can be used to replace the rather cumbersome procedure involving locating sequences.

### 3.11 Method of Das and Farmer

P. Das and D. E. Farmer [54] presented a method for designing fault detection experiments for sequential machines which are realized as parallel connections of simpler component machines. The outputs from these component machines are assumed to be inaccessible for measurement, but it was shown that the knowledge of the machine structure can be utilized to design simpler experiments. The identification of the components of a parallel decomposable sequential machine was carried out by an experiment based on the application of a component distinguishing sequence, the existence of which is less restrictive than the existence of a composite distinguishing sequence. The procedure to deduce the state table of the machine is based upon placing all component machines but one in a fixed reference state prior to the application of the measuring input/output sequences of the said component machine.

Chapter 4

# A Review on the Fault Detection Techniques Developed Through Machine Modification Involving Augmented Logic in Sequential Machines

## 4.1 Introduction

In our discussions in Chapter 3, we observed that the problem of fault detection as well as design of checking sequences for sequential machines is, in general, a rather difficult one to solve. A solution of the problem involves many factors like the total number of states of the machine, whether the machine is *reduced* and *strongly connected*, if the machine possesses distinguishing sequences, and so on. The class of machines that normally possess distinguishing sequences constitutes a very small subclass of the class of all machines. In Chapter 3, it is shown through the works of Hennie and others that the distinguishing sequences can be used to reduce considerably the total length of the checking experiment. In case a sequential machine does not possess distinguishing sequence, the derivation of a checking experiment is still possible but it becomes more complicated. A diagnosable sequential machine is defined as one which possesses one or more distinguishing sequences. In one of the approaches of fault detection in sequential machines developed recently, a given machine can be embedded in a larger machine that possesses distinguishing sequence or sequences. Under such circumstances, the given machine still maintains its original input-output relationship, besides receiving the distinguishing sequence or sequences from the larger machine that contains it. Two basic techniques have been applied for the purpose of constructing an augmented machine of which the given machine can be treated as a *submachine*.* These two techniques are as follows:

(i) The addition of extra input terminals and associated logic to the original machine.

(ii) The addition of extra output terminals and associated logic to the original machine.

*A submachine $M' = (I', S', O', f', g')$ of a machine $M = (I, S, O, f, g)$ is a machine with $I' \subset I$, $S' \subset S$, $O' \subset O$, $f' = f$ restricted to $S' \times I'$, and $g' = g$ restricted to $S' \times I'$.

It has been shown through the works of Kohavi and Lavallee [133], Murakami et al. [160], Holborow [116], Kane and Yau [125], Sheppard and Vranesic [207], Fujiwara and Kinoshita [84], Fujiwara et al. [85] and others that the problem of fault detection and that of designing checking sequences for a sequential machine becomes much simplified when the sequential machine is modified according to any of the above-mentioned techniques. A discussion of the methods that utilize these techniques of machine modification for fault detection and designing checking sequences for sequential machines will be presented in this chapter. These discussions will be useful in the context of the subsequent chapters.

## *4.2 Method of Kohavi and Lavallee

Kohavi and Lavallee [133] suggested a scheme for modifying a given sequential machine in order that the machine might possess certain special distinguishing sequences. They first formed a modified machine containing the original one with the help of some additional output logic. This approach results in checking experiments where the inputs are applied to the terminals of the circuit, and not to any point within the circuit. The extra terminals, however, are predesigned to enable efficient maintenance of the circuit. Given a sequential machine $M$, the objective of their method is to obtain an augmented machine $M'$ containing the original machine $M$, and in addition, possessing some prescribed distinguishing sequences. They defined a definitely diagnosable (DD) machine as a sequential machine having a sequence of length $l$ as a distinguishing sequence, where $l \leqslant q(q-1)/2$, $q$ being the number of states of the machine. The integer $l$ is said to be the *order* of diagnosability. The tools used for achieving this special distinguishing sequence, *viz.*, the testing table and the testing graph we will consider first. An analysis will show that a machine $M$ will become a definitely diagnosable (DD) machine, if and only if, its testing graph $G$ remains loop-free, and no repeated entries exist in the testing table.

To explain the afore-mentioned terms properly, let us consider the example of a machine as shown in Table 4.1. The corresponding testing

Table 4.1 Machine $M_1$

| Present states | Next states and present outputs | |
|---|---|---|
| | $I_1 = 0$ | $I_2 = 1$ |
| $S_1$ | $S_2$, 0 | $S_4$, 0 |
| $S_2$ | $S_1$, 0 | $S_2$, 0 |
| $S_3$ | $S_4$, 0 | $S_1$, 0 |
| $S_4$ | $S_4$, 1 | $S_3$, 0 |

*Illustrations (tables and figures etc.) in section 4.2 are from the paper referred [133] in the bibliography and are presented with the permission of IEEE (copyright © 1967 IEEE).

table is shown in Table 4.2, whereas the testing graph is shown in Fig. 4.1.

Table 4.2 Testing table for machine $M_1$

| | 0/0 | 0/1 | 1/0 | 1/1 |
|---|---|---|---|---|
| $S_1$ | $S_2$ | — | $S_4$ | — |
| $S_2$ | $S_1$ | — | $S_2$ | — |
| $S_3$ | — | $S_4$ | $S_1$ | — |
| $S_4$ | — | $S_4$ | $S_3$ | — |
| $S_1S_2$ | $S_1S_2$ | — | $S_2S_4$ | — |
| $S_1S_3$ | — | — | $S_1S_4$ | — |
| $S_1S_4$ | — | — | $S_3S_4$ | — |
| $S_2S_3$ | — | — | $S_1S_2$ | — |
| $S_2S_4$ | — | — | $S_2S_3$ | — |
| $S_3S_4$ | — | ($S_4S_4$) | $S_1S_3$ | — |

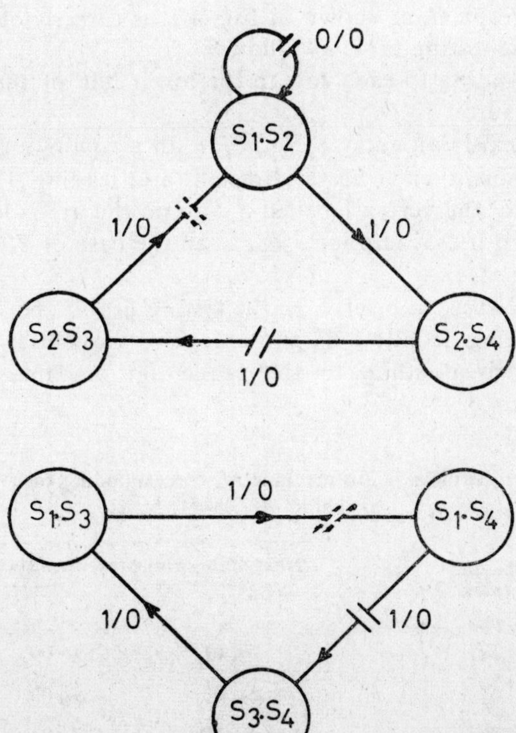


Fig. 4.1 Testing graph for the machine $M_1$.

In the upper half of Table 4.2, the state table of $M_1$ is just rewritten, where (i) the column headings consist of all input-output combinations, and the pair $I_k/O_l$ corresponds to a combination of input $I_k$ and output $O_l$, and (ii) the row headings are the states of the machine. The entry in the column $I_k/O_l$ for row $S_i$ is the $I_k$-successor of $S_i$, if this successor is associated with an output $O_l$, and is a dash (—) otherwise. The lower half of Table 4.2 is derived directly from the upper half. The row headings are all unordered pairs of states, while the table entries are their corresponding successors. If the entries in rows $S_i$ and $S_j$ for column $I_k/O_l$ of the upper half are $S_p$ and $S_q$, respectively, then the entry in row $S_iS_j$ for column $I_k/O_l$ of the lower half is simply $S_pS_q$. If for some pair of states $S_i$, $S_j$, either one or both the entries in column $I_k/O_l$ happens to be dash, the corresponding entry in row $S_iS_j$ for column $I_k/O_l$ is also a dash. Here the pair $(S_iS_j)$ is called an uncertainty pair, whereas its successor $(S_pS_q)$ is called an implied pair. An uncertainty pair that does not imply any other pair, that is, for which all the entries in the corresponding row are dashes, can be omitted from the table. Whenever an entry in the testing table consists of a repeated state (viz., $S_4S_4$ in row $S_3S_4$), that entry is circled. In Table 4.2, the circle around $S_4S_4$ means that states $S_3S_4$ are merged under input 0 to the state $S_4$, and are hence indistinguishable by an experiment which starts with a 0 input.

The testing graph $G$, as shown in Fig. 4.1, is directly obtained from the lower half of the testing table as follows:

(i) Corresponding to each row in the lower half of the testing table, there will be a vertex in $G$.

(ii) If there exists an entry $S_pS_q$, $p \neq q$, in a row $S_iS_j$ for column $I_k/O_l$ of the testing table, then $G$ has a directed arc leading from the vertex labelled $S_iS_j$ to the vertex labelled $S_pS_q$, and the arc is labelled as $I_k/O_l$. No arc is needed if $S_iS_j$ implies $S_pS_p$, as in the case of $S_4S_4$ in row $S_3S_4$ of the example.

Now let us discuss, in brief, how the testing graph and testing table are utilized for making definitely diagnosable (DD) sequential machines. The DD machine corresponding to the sequential machine of Table 4.1 is shown in Table 4.3.

Table 4.3 DD machine $M_1'$ corresponding to machine $M_1$

| Present states | Next states and present outputs | |
|---|---|---|
| | $I_1 = 0$ | $I_2 = 1$ |
| $S_1$ | $S_2$, 01 | $S_4$, 00 |
| $S_2$ | $S_1$, 00 | $S_2$, 00 |
| $S_3$ | $S_4$, 10 | $S_1$, 01 |
| $S_4$ | $S_4$, 11 | $S_3$, 01 |

If the testing table contains a repeated entry in a row $S_iS_j$ for column $I_k/O_l$, then the state $S_i$ cannot be distinguished from the state $S_j$ by an experiment which starts with input $I_k$. Hence to make a machine DD, the undernoted steps are to be followed.

(i) Eliminate all the circled entries by assigning different outputs to the corresponding next state entries.

(ii) Open all the loops of the testing graph by eliminating the smallest number of branches in the graph.

A branch is eliminated by assigning different output symbols to the next state entries which are covered by the node to which it leads. The choice of the branches for elimination is based on two criteria. The first criterion is the reduction of the length of the various paths in the graph, i.e., a branch is chosen for elimination if it opens a loop, and in addition, it opens some paths in the graph. The aim is to open loops, and simultaneously to minimize the length of the longest path in the graph. The second criterion is the minimization of the number of eliminated branches. The aforesaid procedure can now be applied to the said machine $M_1$ in the following way.

(i) The elimination of the repeated entries in the testing table is done by assigning an output 10 to the transition from $S_3$ to $S_4$, and an output 11 to the transition from $S_4$ to $S_4$.

(ii) The self-loop around $S_1S_2$ is opened by assigning the output symbols 01 and 00, respectively, to the next state entries $S_2$ and $S_1$ in column $I_1 = 0$. The loop $S_1S_2 - S_2S_4 - S_2S_3 - S_1S_2$ can be opened by eliminating the arc leading from $S_2S_4$ to $S_2S_3$, and this is achieved by assigning the output symbols 00 and 01 to the next state entries $S_2$ and $S_3$ in rows $S_2$ and $S_4$ for column $I_2 = 1$, respectively. In a similar manner, the loop $S_1S_3 - S_1S_4 - S_3S_4 - S_1S_3$ can be removed by eliminating the arc from $S_1S_4$ to $S_3S_4$, by assigning a 00 output to the next state entry $S_4$ in row $S_1$ for column $I_2 = 1$.

All such assignments of different output symbols to selected transitions were done by augmenting the machine with the help of an extra output terminal as shown in Fig. 4.2. After these steps have been carried out, an inspection shows that the output associated with the state $S_3$ in column

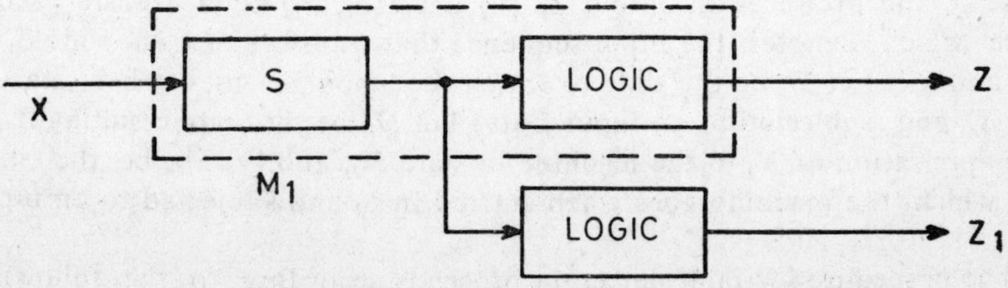


Fig. 4.2 Block diagram of the DD machine $M_1'$ that corresponds to the original machine $M_1$, where $Z_1$ is used for diagnosing purposes only.

$I_2 = 1$ is still not specified. The testing graph reveals that the longest path contains two branches and hence the DS's are, at most, of length three. It is discussed in [133] that if the testing table of the machine $M_1$ is free of repeated entries, the testing graph is loop-free, and if the length of the longest path in the graph is $\mu$, then $l$, the order of diagnosability, becomes $l = \mu + 1$. Thus, in addition, let the output entry of the state $S_3$ for input $I_2 = 1$ be taken as 01 so that it may reduce the lengths of the paths in the testing graph, and the distinguishing sequences may become, at most, of length two. This specification of the output as 01 cancels the branches from $S_1S_3$ to $S_1S_4$, and from $S_2S_3$ to $S_1S_2$ in the testing graph.

From the above discussion, it becomes apparent that even if a sequential machine does not possess any DS, the said machine can be converted to a DD machine. The DD machine has the remarkable advantage of possessing many distinguishing sequences so that the experimenter designing fault detection experiments for the machine may choose the most convenient ones. It is, however, extremely difficult to use more than one DS in a single experiment. Although every sequence of length $l$ is a DS for a properly operated DD machine, it is rather next to impossible to use, in an efficient manner, all or a subset of the set of $2^l$ possible sequences simultaneously in the same experiment, since every state must be uniquely identified, at least once, with the same DS. A method is also presented by Kohavi and Lavallee for the construction of simple, short, and efficient fault detection experiments involving the use of DS's with repeated symbols like 000, or 111, etc. The DD machine has been shown to be a special case of this more general class of machines providing such sequences. The main motivation behind making a machine DD is that there is a need to design fault location experiments, i.e., experiments that not only will detect faults in the machine, but will also indicate the exact location and nature of such faults. It is simpler to design such experiments in the case of DD machines, where every sequence of length $l$ is a DS, and the experimenter can crosscheck the machine with every sequence, not simply with a single one.

To discuss the procedure of fault detection in detail, let us consider (i) $X_0$ a DS of length $L$, for a sequential machine $M$. When the machine $M$ remains in state $S_i$ and the input sequence $X_0$ is applied, it goes to a state $Q_i$ and produces an output $Z_i$. (ii) Let $T(S_i, S_j)$ be a transfer sequence which denotes the input sequence that transfers $M$ from state $S_i$ to $S_j$, and (iii) Let $S_{ij}$ be the state to which the machine goes when started in $S_i$ and subjected to an input $j$. (iv) Let $Q_{ij}$ be the state resulting from the application of $X_0$ to the machine in state $S_{ij}$, and (v) $S_{ijk}$ be the state to which the machine goes when started in $S_i$ and subjected to an input $jk$.

The procedure for fault detection proceeds according to the following steps:

(a) Apply the homing sequence (HS) and transfer sequence $T(S_i, S_0)$ to bring the machine to the starting state $S_0$.

(b) Choose $X_0$ to be the shorter one of the sequences consisting of all 1's or all 0's (for a DD machine, both would be of same length).

(c) Apply $X_0$ followed by a 1 to check the transition from $S_0$ under an input 1. (If an all-zero DS has been chosen, apply an input of 0 instead of 1.)

(d) If $S_{01} \neq S_0$, apply another input 1 to check the transition from $S_{01}$ under an input 1. This checks the transition from $S_{01}$ under an input of 1. Similarly, if $S_{011} \neq S_{01}$, and $S_{011} \neq S_0$, apply another 1. In the same manner, continue with adding 1 inputs as long as new transitions are being checked.

(e) When an additional 1 input does not yield any new transition, apply an input of 0 followed by $X_0$.

(f) As long as new transitions can be checked, apply inputs of 1 to the machine. When no new transitions can be checked, repeat (e) and (f).

(g) When (e) and (f) do not yield any new transitions, and the machine, which is in state $S_i$, is not yet completely checked, apply $T(S_i, S_k)$, where $S_k$ is the state whose transition has not been checked, such that $T(S_i, S_k)$ passes through already checked transitions only.

(h) Repeat (e), (f), (g) until all transitions have been checked.

Let us now apply the above procedure to the DD machine $M_1'$ as shown in Table 4.3. In order to simplify the notation, let us use the decimal value of the output, i.e., 10 = 2, 11 = 3, etc. Let us choose $X_0 = 11$. The machine can be brought to the initial state $S_1$ before the experiment starts by the application of a suitable homing sequence followed by a transfer sequence, since every sequence of length two is also a homing sequence for the machine. The complete checking sequence developed with the machine $M_1'$ in the starting initial state $S_1$ following the aforesaid procedure is given below.

Input : 1 1 1 1 1 0 1 1 0 1 1

State : $S_1$ $S_4$ $S_3$ $S_1$ $S_4$ $S_3$ $S_4$ $S_3$ $S_1$ $S_2$ $S_2$ $S_2$

Output: 0 1 1 0 1 2 1 1 1 0 0

Input : 1 0 1 1 0 0 1 1

State : $S_2$ $S_1$ $S_4$ $S_3$ $S_4$ $S_4$ $S_3$

Output: 0 0 0 1 2 3 1 1

This checking sequence is not the only checking sequence for the machine $M_1'$ of Table 4.3, but it is unique for $M_1'$ only being started in state $S_1$. There is no other machine (except that isomorphic to $M_1'$) that possesses the same checking sequence. This is because (i) the fault detection experiment clearly demonstrates that the machine under study contains at least four distinct states, (ii) every transition is followed by DS, which ascertains the transition, and uniquely identifies the state to which the machine goes, in terms of the previously defined four states.

The DS is applied $2q + 1$ times (in this case nine times), and identifies $2q$ transitions as well as the final state. Therefore, as the machine is deterministic, the above experiment corresponds to that on a 4-state machine isomorphic to $M_1'$.

An upper bound on the length of the checking sequence for a $q$-state machine with $p$ input symbols can be found based on the procedure of the experiment as stated earlier. The bound $\xi$ can be shown to be less than or equal to

$$pq + q(p-1)L_{ds} + L_{ds} + (p-1)(q-1)^2$$

Since $L_{ds}$, the length of the DS $\leqslant q(q-1)/2$, therefore, this bound $\xi$ comes to the order of $pq^3$.

The above method of fault detection remains valid as long as we assume that the faults that occur never result in an increase in the number of states of the machine. Indeed, if we assume a completely specified machine having $q = 2^m$ states, $m$ being the number of state variables, then the possibility of occurrence of such faults is rather small, since this theoretically implies the creation of new state variables. A minimum number of additional outputs may not always result in the process of making a machine definitely diagnosable, if a minimum number of branches are opened in the testing graph. Consequently, a certain amount of 'trial and error' becomes unavoidable in order to achieve the goal of minimum extra output variables. The problem of providing an algorithm for finding the required minimum number of additional outputs is very intricate and still appears unsolved. In order to profitably utilize the properties of the DD machines and the numerous distinguishing sequences available, it is felt, however, that more efforts are needed to study the properties of these machines and their various distinguishing sequences to get a better insight. It has been claimed in [133] that the utilization of the additional logic as used in the machine does not appear too costly when looked into from the viewpoint of reducing the length of the complex and lengthy experiments.

## *4.3 Counter cycle approach of Murakami et al.

Soon after Kohavi and Lavallee [133] developed their approach of fault detection with the aid of an augmented logic, S.-I. Murakami, K. Kinoshita and Z. Ozaki [160] demonstrated that the problem of designing checking sequences for sequential machines including their fault diagnosis can be made much easier if a given machine can be modified by an extra input symbol thereby making the machine a counter cycle (CC) type machine. The counter cycle and the CC machine are defined in Section 2.14 of Chapter 2. It can be recalled that a counter cycle $C_I$ of a $q$-state sequential machine is an alternating sequence of state $S_i$ and input symbol $I$ such that

*Illustrations (tables etc.) in Section 4.3 are from paper referred [160] in the bibliography and are presented with the permission of IEEE (copyright © 1970 IEEE).

$$f(S_i, I) = S_{i+1} \text{ and } g(S_i, I) = 0,\ 1 \leqslant i \leqslant q - 1$$

and

$$f(S_q, I) = S_1 \text{ and } g(S_q, I) = 1,$$

where $f$ and $g$ denote, respectively, the next state and output mappings of the machine.

We have seen earlier in Chapter 3 that the entire process of fault detection becomes difficult for sequential machines which are not strongly connected. Murakami et al. introduced an augmentation scheme to a given sequential machine with the help of an extra input symbol $I$ satisfying a counter cycle, and demonstrated that the modified machine not only becomes always strongly connected but also contains the original one and simplifies to a great extent the entire process of fault detection in the machine. The scheme of the fault detection suggested by Murakami et al. reduces the length of the checking sequence. Here for a $q$-state machine with $p$ input symbols, the authors [160] obtained a length of the checking sequence that does not exceed $2q + 2pq^2$ symbols. Clearly, the order of this length is $pq^2$ symbols, which is less compared to that obtained by Kohavi and Lavallee [133], where the order comes as $pq^3$ symbols.

In the context of a CC machine, the following input/output sequences were defined.

(a) $D_i$ *sequence*: For a $q$-state CC machine with a counter cycle $C_I$, a $D_i$ sequence for a state $S_i$ is defined as an input/output sequence such that (i) the input sequence consists of $(q - i + 1)$ consecutive symbols $I$, and (ii) the output sequence is a concatenation of a sequence comprised of $(q - i)$ consecutive output symbols 0, and only one output symbol 1.

(b) $T_i$ sequence: For a $q$-state CC machine with counter cycle $C_I$, a $T_i$ sequence of a state $S_i$ is stated to be an input/output sequence such that (i) the input sequence consists of $(i - 1)$ consecutive input symbols $I$, while (ii) the output sequence is $(i - 1)$ consecutive output symbols $O$.

(c) $\alpha_q$ *sequence (for a q-state machine)*: For a $q$-state CC machine with counter cycle $C_I$, an $\alpha_q$ sequence is defined as an input/output sequence of length $2q - 1$ symbols such that (i) the input sequence consists of $2q - 1$ input symbols $I$, and (ii) the output sequence is a concatenation of a sequence of $q - 1$ consecutive symbols 0, a symbol 1, and again a sequence of $q - 1$ consecutive symbols 0.

(d) $\beta_q$ *sequence (for a q-state machine)*: A single input symbol $I$ with its corresponding output symbol 1 preceded by the $\alpha_q$ sequence of the CC machine having counter cycle $C_I$ forms a $\beta_q$ sequence of the said machine. Thus $\beta_q = \alpha_q\binom{I}{1}$. If $\beta_q$ sequence is applied to such machine in the initial state $S_1$, then its application will take the machine through all the possible states twice and ultimately set the machine back to the state $S_1$. Of course, it implies that the $\beta_q$ sequence remains valid as long as the number of states of the machine are not increased due to any kind of faults.

(e) $\gamma_{ij}$ *sequence*: If the transition of the machine under the input $I_j$

happens between the states $S_i$ and $S_k$ producing an output $O_j$, then $\gamma_{ij}$ sequence is defined as follows:

$$\gamma_{ij} = (T_i)\binom{I_j}{O_j}(D_k)$$

Here the $T_i$ sequence preceding the said input/output pair sets the machine to the state $S_i$ and the $D_k$ sequence thus followed definitely identifies the state $S_k$ and also returns the machine at the end to the initial state $S_1$.

Thus concatenating with the $\beta_q$-sequence all the $\gamma_{ij}$ sequences ($1 \leqslant i \leqslant q$ and $1 \leqslant j \leqslant p$) for a $q$-state machine with $p$-input symbols, the required checking sequence of the CC machine can be obtained.

In all practical situations, there are very few machines that exhibit counter cycle for a particular input symbol. The above-mentioned approach of fault detection and that of designing checking sequence given by Murakami et al. remains valid to such machines if they can be predesigned with an extra input symbol $I_e$, such that the modified machine contains the original one and also serve the purpose of CC machine having counter cycle in input symbol $I_e$. This can be done by defining the next state and output functions for this special augmented input symbol $I_e$ as follows:

$$f(S_i, I_e) = S_{i+1}; \quad g(S_i, I_e) = 0, \quad 1 \leqslant i \leqslant q-1$$

$$f(S_q, I_e) = S_1; \quad g(S_q, I_e) = 1$$

where $f$ and $g$ represent the next state and output functions of the machine, respectively. It is obvious that the modified machine thus obtained will be a CC machine with a counter cycle $C_{I_e}$.

*Example*: Consider the machine $M_2$ as shown in Table 4.4, its corresponding modified machine $M_2'$ can be obtained and is shown in Table 4.5.

Table 4.4 Machine $M_2$

| Present states | Next states and present outputs | |
|---|---|---|
| | $I_1 = 0$ | $I_2 = 1$ |
| $S_1$ | $S_3$, 0 | $S_3$, 1 |
| $S_2$ | $S_3$, 0 | $S_1$, 0 |
| $S_3$ | $S_2$, 1 | $S_3$, 1 |

Table 4.5 Modified machine $M_2'$

| Present states | Next states and present outputs | | |
|---|---|---|---|
| | $I_1 = 0$ | $I_2 = 1$ | $I_3 = I_e$ |
| $S_1$ | $S_3$, 0 | $S_3$, 1 | $S_2$, 0 |
| $S_2$ | $S_3$, 0 | $S_1$, 0 | $S_3$, 0 |
| $S_3$ | $S_2$, 1 | $S_3$, 1 | $S_1$, 1 |

The amount of hardware required to introduce this new input symbol $I_e$ is also estimated by Murakami et al. as follows:

(i) no extra memory elements are required,
(ii) some additional combinational logic which determines state transitions for a new input symbol $I_e$ is required,
(iii) one additional input terminal is required.

Of course, the hardware requirement of (ii) and (iii) mentioned above may be brought down to some extent by choosing an appropriate input variable assignment.

Let us now illustrate the designing of fault detection experiment utilizing the above concepts for the machine $M_2'$ as shown in Table 4.5.

In this case the $\beta_q$ sequence is of the form

$$\beta_3 = \begin{pmatrix} \text{Input sequence} \\ \text{Output sequence} \end{pmatrix} = \begin{pmatrix} I_e & I_e & I_e & I_e & I_e & I_e \\ 0 & 0 & 1 & 0 & 0 & 1 \end{pmatrix}$$

The $D_i$ and $T_i$ sequences corresponding to every state $S_i$ of the modified machine $M_2'$ is shown in Table 4.6, whereas the respective $\gamma_{ij}$ sequences are shown in Table 4.7.

Table 4.6 $D_i$ and]$T_i$ sequences for $M_2'$

| States | $D_i$ sequence | $T_i$ sequence |
|---|---|---|
| $S_1$ | $\begin{pmatrix} I_e & I_e & I_e \\ 0 & 0 & 1 \end{pmatrix}$ | $\lambda$ (null) |
| $S_2$ | $\begin{pmatrix} I_e & I_e \\ 0 & 1 \end{pmatrix}$ | $\begin{pmatrix} I_0 \\ 0 \end{pmatrix}$ |
| $S_3$ | $\begin{pmatrix} I_e \\ 1 \end{pmatrix}$ | $\begin{pmatrix} I_e & I_e \\ 0 & 0 \end{pmatrix}$ |

Table 4.7 $\gamma_{ij}$ sequences for $M_2'$

| States | $I_1 = 0$ | $I_2 = 1$ |
|---|---|---|
| $S_1$ | $\begin{pmatrix} 0 \\ 0 \end{pmatrix}\begin{pmatrix} I_e \\ 1 \end{pmatrix}$ | $\begin{pmatrix} 1 \\ 1 \end{pmatrix}\begin{pmatrix} I_e \\ 1 \end{pmatrix}$ |
| $S_2$ | $\begin{pmatrix} I_e \\ 0 \end{pmatrix}\begin{pmatrix} 0 \\ 0 \end{pmatrix}\begin{pmatrix} I_e \\ 1 \end{pmatrix}$ | $\begin{pmatrix} I_e \\ 0 \end{pmatrix}\begin{pmatrix} 1 \\ 0 \end{pmatrix}\begin{pmatrix} I_e & I_e & I_e \\ 0 & 0 & 1 \end{pmatrix}$ |
| $S_3$ | $\begin{pmatrix} I_e & I_e \\ 0 & 0 \end{pmatrix}\begin{pmatrix} 0 \\ 1 \end{pmatrix}\begin{pmatrix} I_e & I_e \\ 0 & 1 \end{pmatrix}$ | $\begin{pmatrix} I_e & I_e \\ 0 & 0 \end{pmatrix}\begin{pmatrix} 1 \\ 1 \end{pmatrix}\begin{pmatrix} I_e \\ 1 \end{pmatrix}$ |

By concatenating $\beta_3$ sequence with all $\gamma_{ij}$ sequences obtained from Table 4.7, the checking sequence of the machine $M_2'$ can be obtained as follows:

$$\begin{pmatrix} I_e & I_e & I_e & I_e & I_e & I_e \\ 0 & 0 & 1 & 0 & 0 & 1 \end{pmatrix}\begin{pmatrix} 0 & I_e \\ 0 & 1 \end{pmatrix}\begin{pmatrix} I_e & 0 & I_e \\ 0 & 0 & 1 \end{pmatrix}\begin{pmatrix} I_e & I_e & 0 & I_e & I_e \\ 0 & 0 & 1 & 0 & 1 \end{pmatrix}$$

$$\begin{pmatrix} 1 & I_e \\ 1 & 1 \end{pmatrix}\begin{pmatrix} I_e & 1 & I_e & I_e & I_e \\ 0 & 0 & 0 & 0 & 1 \end{pmatrix}\begin{pmatrix} I_e & I_e & 1 & I_e \\ 0 & 0 & 1 & 1 \end{pmatrix}$$

This checking sequence does remain applicable under some valid assumptions regarding the machine's failures. The following assumptions were made in respect of the nature of circuit faults in the study of Murakami et al.:

(i) When failure occurs in a sequential machine, it simply operates as another sequential machine.

(ii) The failures occurring in a machine do not increase the number of states of the machine.

(iii) No new failures do occur during the fault detection experiments.

So far, we have discussed the procedure of construction of the checking sequences for the CC machines. Murakami et al. also developed an approach for the fault diagnosis of a sequential machine possessing counter cycles. An adaptive experimental procedure was developed for finding the state table of the (faulty) machine on the assumption that no failures occur in any state transition by the extra input symbol $I_e$. This procedure can be depicted as follows:

*Step 1*: Apply the input $I_e$ repeatedly to the given machine at most $q$ times until an output of 1 is obtained. If the output 1 is not obtained due to the application of $q$ repeated input symbols $I_e$, this indicates that a failure has occurred not satisfying the original basic assumptions.

*Step 2*: Apply the inputs of the $\beta_q$ sequence, and compare the output sequence of the original machine with that of the $\beta_q$ sequence. If the required response is obtained, then go to Step 3. Otherwise, a failure is indicated for the input symbol $I_e$.

*Step* 3: The state transitions are obtained as follows:

(a) Apply the $T_i$ sequence to the machine.

(b) Apply an input $I_j$ to the machine. (Note if the machine produces an output $O_j$ for the input $I_j$.) This takes the machine from the state $S_i$ to some other state $S_x$ with an output $O_j$ in response to the input $I_j$.

(c) Apply the input $I_e$ repeatedly to the machine until an output of 1 is obtained. If an output 1 is obtained at the $k$th time, then we decide that the next state of $S_i$ for the input $I_j$ is $S_x = S_{q-k+1}$. This way all possible state transitions can be obtained.

The afore-mentioned concepts of fault diagnosis and that of designing checking sequences for sequential machines as developed by Murakami et al., were based upon the assumption of nonfaulty performance of the machine for the special input symbol. No alternative to the application of a single special input symbol satisfying counter cycle was taken into consideration. The probability of the failures in the said input could be minimized by applying two inputs, both satisfying counter cycles. This possibility will be discussed later in Chapter 5. Murakami et al. treated the case for Mealy model machines only, and no suggestions were made to the extension of the ideas in the case of Moore model sequential machines. Next, the basic assumption upon which the entire fault detection experiment depends, viz., that no failures will occur which may increase the number of states of the machine seems to be of limited significance. At the end of their discussions, the authors indicated the possibility of schemes for the construction of other types of machines, viz., machines in which (i) a sequence of $n$ consecutive special input symbols (for an $n$-state machine) represents a synchronizing sequence, or (ii) a special input symbol $I$ represents a distinguishing input symbol. Though it was stated that the procedures of constructing transfer sequences in these two cases may not be as systematic as in the case of CC machines, yet, no further comments regarding the advantages of the construction of such machines over that of the CC machines were mentioned.

## 4.4 Improvement on the techniques of Murakami et al.

C. E. Holborow [116] improved the bound on the length of the checking experiments as derived by Murakami et al. [160] by assigning a more efficient output specification in the counter cycle. The method consists of designing in a $q$-state machine a binary sequence of period $q$, such that each of the $q$ subsequences of length $m = \lceil \log_2 q \rceil$ becomes distinct. Here again, the ceiling operator denotes the smallest integer greater than or equal to the number inside it. The outputs of the machine under the special input symbol are so assigned that the machine produces the assumed sequences of period $q$ when it is in the counter cycle. Thus, if the machine starts in a state $S_i$, the $m$ outputs it produces in response to the $m$ applications of the special input symbol satisfying the counter cycle identifies the state $S_i$ uniquely. The $m$-bit sequence required for this purpose is generated by a $m$-bit shift register.

The checking sequence consists of (i) a subsequence of length $q + m$ to verify that the counter cycle is functioning properly, and (ii) at most $pq$ subsequences to check state transitions of the machine, where $p$ denotes the number of input symbols in the input alphabet of the original machine. Each transition is checked by a subsequence consisting of (i) a transfer sequence of length at most $q - 1$ to take the machine to the desired state, (ii) a sequence consisting of a single input to perform the transition under verification, and (iii) a distinguishing sequence of length $m$ to verify that the transition ended in the correct state. The procedure gives an upper bound on the length of the checking sequence as follows:

$$q + m + pq \cdot (q + m) = (q + \lceil \log_2 q \rceil)(1 + pq)$$

For large values of $q$, this bound is slightly in excess of half the bound as obtained by Murakami et al. [160].

So far we have discussed how Murakami et al. [160] modified the input logic by adding a permutation input that can provide a homogeneous distinguishing sequence. Kane and Yau [125] later improved the method with a more suitable output assignment under the permutation input. They also considered using additional inputs to provide transitions needed to transform the cell graph into an Eulerian circuit. Note that a cell is defined exactly similarly as in the method of Gönenc [96], and the state table and the graph associated with such cells are referred to as the cell table and cell graph, respectively. Taking three assumptions to be valid, that is, (i) the faults do not occur during the test, (ii) they last at least during the test, and (iii) the state count never increases as a result of the faults, Sheppard and Vranesic [207] further improved this method of fault detection in a completely specified synchronous sequential machine utilising the graph-theoretic technique of Gönenc [96]. This technique was algorithmic, and was based on the concept of embedding a given binary machine into an easily testable $R$-valued machine ($M'$), where $R \geqslant 2$. The test sequence was developed for the latter, and the resultant machine was then implemented either in many-valued or in some other encoded form. It was shown that a heuristic optimization of additional permutation inputs reduces the length of the checking sequences considerably.

A bound on the length of the checking sequences was also derived in [207]. For a $q$-state machine with $p$ input symbols, this bound is less than or equal to

$$2q + L_{ds} + p(L_{ds} + 1)(2q - L_{ds} - 2) + p \sum_{i=1}^{L_{ds}+1} i,$$

where $L_{ds}$ is the length of the distinguishing sequence.

There are of course two disadvantages of this technique. First, the properties of the given machine are not always employed efficiently. Thus the possibility of using transfer sequences from the given machine is ignored as well as the use of multiple overlapping distinguishing sequences. Secondly, the augmentation of input and output logic may require a number of additional leads, particularly in the binary-coded implementation. The limitation becomes more severe in the case of single-input single-output machines with a large number of states. But this provides indirectly an incentive for many-valued logic implementation.

## *4.5 Output-observable machines for fault diagnosis

As a contemporary work of Sheppard and Vranesic, H. Fujiwara and K. Kinoshita [84] have introduced an output-observable sequential machine for the purpose of designing checking sequences. For a $k$-output observable sequential machine, one can find a checking sequence $w_1 w_2$ such

*Illustrations (tables etc.) in Section 4.5 are from the paper referred [84] in the bibliography and are presented with the permission of IEEE (copyright © 1974 IEEE).

that $w_1$ is an input-output sequence of which the input sequence takes a given state table through all the transitions, and $w_2$ is an arbitrary input-output sequence of length $k$. Since a checking sequence must take the machine through all the state transitions, the length of the checking sequence $w_1w_2$ is nearly minimum. The output-observable sequential machine has been discussed in Section 2.15 of the Chapter 2. A procedure is given in [84] for the modification of a given machine to its output-observable version with minimum number of extra outputs. In Chapter 2, in Table 2.6, the machine as shown is an 1, 2 output-observable with respect to $z_1 \times z_2$ and the zero-partition. For the same machine, by applying an arbitrary input sequence of length $k$ ($k = \max\{1, 2\} = 2$), and observing the output sequences of length 1 and 2 at the output terminals $z_1$ and $z_2$, respectively, one can establish the initial and final state of the machine. Suppose that the machine of Table 2.6 is in state $S_1$; then the shortest input-output sequence $w_1$ to take the machine through all the state transitions is obtained as follows:

| | | | | | | | | | | | |
|---|---|---|---|---|---|---|---|---|---|---|---|
| Input : | | 0 | 0 | 0 | 1 | 1 | 1 | 0 | 1 | 0 | 1 |
| State : | $S_1$ | $S_4$ | $S_5$ | $S_5$ | $S_1$ | $S_2$ | $S_2$ | $S_3$ | $S_3$ | $S_4$ | $S_5$ |
| Output $z_2$: | | 1 | 0 | 1 | 1 | 1 | 0 | 0 | 0 | 0 | 0 |
| Output $z_1$: | | 0 | 1 | 1 | 1 | 0 | 0 | 0 | 1 | 1 | 1 |

The following input-output sequence of length 2 represents $w_2$ starting at state $S_5$:

| | | | |
|---|---|---|---|
| Input : | | 0 | 0 |
| State : | $S_5$ | $S_5$ | $S_5$ |
| Output $z_2$: | | 1 | 1 |
| Output $z_1$: | | 1 | 1 |

The checking sequence of the given machine represents the concatenation of the above two input-output sequences, and consists of twelve input symbols, ten for $w_1$, and two for $w_2$.

The assumption in this method that the faulty machine, irrespective of all types of faults, will not cease to show the criterion for output-observability points to the limitation of this technique. In the technique the minimum number of extra output variables that will be required is specified, whereas it can be recalled that a certain amount of 'trial and error' was necessary in the method of Kohavi and Lavallee [133] to have the minimum extra output variables for making a DD machine. As in the case of other methods, this method also fails to give a correct result for the checking sequence, if a fault increases the number of states of the machine. One major advantage of the method lies in the fact that the checking sequence is required to take the machine through all state transitions only, and hence it can be tested rather easily.

## *4.6 Augmented easily testable machines

In a subsequent work following the one as mentioned in [84], H. Fujiwara, Y. Nagao, T. Sasao and K. Kinoshita [85] described a scheme of modification of an arbitrary sequential machine for the purpose of easily testing it with the help of two extra input symbols. A $q$-state machine is said to be an easily testable machine, if and only if it possesses (i) a distinguishing sequence $X_d$ of length $\lceil \log_2 q \rceil$ which forces the machine into a specific state $S_1$, and (ii) a transfer sequence $T_i$ of length at most $\lceil \log_2 q \rceil$ to take the machine from state $S_1$ to a state $S_i$, for all $i$. For a $q$-state sequential machine with $p$ input symbols, the procedure of Fujiwara et al. [88] gives a satisfactory bound on the length of the checking sequence which is approximately $pq \cdot \lceil \log_2 q \rceil$ symbols.

The augmented machine needs the following:

(i) The addition of new states to the original machine $M$ is done if its number of states $q$ is not an integral power of 2. Thus, $S_{q+1}, S_{q+2}, \ldots, S_{q'}$ states are added to $M$ where $q' = 2^m$ and $m = \lceil \log_2 q \rceil$.

(ii) An $m$-bit binary code is assigned to all such states so that each state is given one assignment only.

(iii) Two new input symbols $I_{e1}$ and $I_{e2}$ are added to $M$. The next state function $f$, and the output function $g$, for the two new input symbols $I_{e1}$ and $I_{e2}$ are defined as follows:

For each state $S_i$ with a state assignment $Y_1 Y_2 \ldots Y_m$,

$$f(S_i, I_{e1}) = S_j; \quad f(S_i, I_{e2}) = S_k;$$

and

$$g(S_i, I_{e1}) = g(S_i, I_{e2}) = 0_1, \text{ if } Y_m = 0;$$
$$= 0_2, \text{ if } Y_m = 1;$$

where the states $S_j$ and $S_k$ have the state assignments $0Y_1Y_2 \ldots Y_{m-1}$ and $1Y_1Y_2 \ldots Y_{m-1}$, respectively. The effect of this transition makes this two-column submachine restricted to the input symbols $I_{e1}$ and $I_{e2}$ isomorphic to an $m$-stage binary shift register.

As an example, let us consider the machine $M_3$ as shown in Table 4.8. Its augmented easily testable version $M_3'$ is shown in Table 4.9.

Table 4.8 Machine $M_3$

| Present states | Next states and present outputs | |
|---|---|---|
| | $I_1 = 0$ | $I_2 = 1$ |
| $(00)S_1$ | $S_2$, 1 | $S_1$, 1 |
| $(01)S_2$ | — | $S_3$, 0 |
| $(10)S_3$ | $S_2$, 0 | —, 1 |

*Illustrations (tables etc.) in section 4.6 are from the paper referred [85] in the bibliography and are presented with the permission of IEEE (copyright © 1975 IEEE).

Table 4.9 Augmented machine $M_3'$

| Present states | Next states and present outputs | | | |
|---|---|---|---|---|
| | $I_1 = 0$ | $I_2 = 1$ | $I_{e1}$ | $I_{e2}$ |
| $(00)S_1$ | $S_2$, 1 | $S_1$, 1 | $S_1$, 0 | $S_3$, 0 |
| $(01)S_2$ | — | $S_3$, 0 | $S_1$, 1 | $S_3$, 1 |
| $(10)S_3$ | $S_2$, 0 | —, 1 | $S_2$, 0 | $S_4$, 0 |
| $(11)S_4$ | — | — | $S_2$, 1 | $S_4$, 1 |

The entire checking sequence depends upon the validity of two basic assumptions. These are: (i) any failure which occurs is assumed to occur throughout the test, and (ii) no failure will increase the number of states of the machine. The whole checking experiment is a preset one and requires no additional adaptive initializing sequences which bring the machine under test to the starting state. The checking experiment can be organized according to the following five steps:

(i) The application of an initializing part which brings the machine under test to the starting state $S_1$ for the experiment.

(ii) The second part deals with the application of the input sequences which make the correctly operating machine to traverse through all its states and displays the responses to the distinguishing sequence $X_d$. Obviously, this also in turn verifies that $X_d$ is a distinguishing sequence.

(iii) The third part consists of the application of $X_d$ again to verify that $X_d$ represents a synchronizing sequence also.

(iv) The fourth part deals with the application of input sequences to verify that $T(i)$ represents the transfer sequence used to transfer the machine from the state $S_1$ to a state $S_i$, for all $i$.

(v) The last part consists of applying the input symbols to check all the state transitions.

The above-mentioned steps (i) through (iv) can be combined as follows:

| | | | | | |
|---|---|---|---|---|---|
| Input : | | $X_d$ | $T(i)$ | $X_d$ | $X_d$ ; |
| State : | — | $S_1$ | $S_i$ | $S_1$ | $S_1$; |
| Output: | | — | $Z_{1i}$ | $Z_i$ | $Z_1$ ; |

for all $i$, where $Z_i = g(S_i, X_d)$, and $Z_{1i} = g(S_1, T(i))$. The last step will give the following input-output sequence, viz.,

| | | | | | | |
|---|---|---|---|---|---|---|
| Input : | | $X_d$ | $T(i)$ | $I_j$ | $X_d$ | |
| State : | — | $S_1$ | $S_i$ | | $S_{ij} = f(S_i, I_j)$ | $S_1$ |
| Output: | | — | $Z_{1i}$ | $O_{ij} = g(S_i, I_j)$ | | $Z_{ij}$ |

This kind of formation of the input-output sequences gives a total length of the checking sequence which is at most equal to

$$|X_d| + \sum_{i=1}^{q} (|T(i)| + 2|X_d|) + \sum_{j=1}^{p} \sum_{i=1}^{q} (|T(i)| + |I_j| + |X_d|)$$

$$= (2q + 1)|X_d| + \sum_{i=1}^{q} |T(i)| + pq(|X_d| + 1) + p \sum_{i=1}^{q} |T(i)|$$

$$\leqslant (2q + 1)\lceil \log_2 q \rceil + q\lceil \log_2 q \rceil + pq(\lceil \log_2 q \rceil + 1) + pq\lceil \log_2 q \rceil$$

$$= (3q + 1)\lceil \log_2 q \rceil + pq(2\lceil \log_2 q \rceil + 1)$$

For large $p$ and $q$, this bound is found to be smaller than the bound $(q + \lceil \log_2 q \rceil) + (1 + pq)$ as reported by Holborow [116].

One limitation in all the above discussed methods of fault detection is the assumption that the faults will never cause an increase in the number of states of the machine, particularly when the number of states $q$ satisfies the inequality $2^{m-1} < q < q' = 2^m$. No scheme has been suggested to make an easily testable machine, taking into consideration the aforesaid aspect of the problem. In Chapter 5, the case of a very easily testable-cum-diagnosable sequential machine will be considered that takes into account the above aspect of the problem. Also, a further modification and improvement of the method of Fujiwara et al. [85] will be considered in Chapter 5.

## 4.7 Shift register modifications for synchronous circuits

It has been shown by M. J. Y. Williams and J. B. Angell [226] that most problems in sequential circuit testing can be overcome if (i) the circuit can easily be set to any desired internal state, and (ii) it is easy to find a sequence of input patterns such that the resulting output sequence will indicate whether the circuit was in a given state. The (ii) item signifies, of course, the distinguishing sequence requirement. The authors could achieve the above-mentioned two properties required for readily testing of the sequential machines by incorporating some modifications in the design of such machines. Their suggestions are based on the principle that the bistables in the circuit can, under the control of a signal, be connected together in a chain to behave as a shift register. This facility can be modelled by a double throw switch in each input lead of every bistable and in one or two of the circuit's primary output connections. All these switches are ganged together and the circuit can operate either in its 'normal' mode or 'shift register' mode. In the shift register mode, the first bistable can be set directly from one of the primary inputs, and the output of the last bistable can be directly monitored on one of the primary outputs. This means that the circuit can be set to any desired state from the primary inputs and that the internal state can be determined by the signal source appearing on the primary outputs. The procedure for testing the circuit is as follows:

*Step 1*: Switch to shift-register mode and check the shift-register operation.

*Step 2*: Set the initial state into the shift register.

*Step 3*: Return to normal mode and apply the test-input pattern.

*Step 4*: Switch to shift-register mode and shift out the final state while setting the starting state for the next test. Return to Step 3.

## 4.8 Use of control logic to modify sequential machines to simplify test generation

With a view to designing systems requiring relatively few tests for fault detection and location, J. P. Hayes [102] explored the possibility of using control logic to reduce the number of tests required by a logic network to simplify test generation.

In sequential machines, the problem of identifying the states of a machine during testing has a significant effect on the test length. It is shown by Hayes that a control logic can be added to the machine to make it easier to determine the states of the machine, or to help resetting the machine to a known state; of course, in the latter case, the reset circuit may not be testable. In his approach, the author also examined what types of control logic should be suitable, and also to which part of the circuit the control has to be placed in the implementation of the circuit. Hayes [102] treated such problems by using EOR module as control element, and suggested modifications in designing the original sequential machine itself. The resulting circuit consists of alternating EOR modules and NAND gates, and requires only five tests for checking.

Chapter 5

# Fault Detection in Sequential Machine Utilizing Extra Inputs and Extra Outputs

## 5.1 Introduction

In Chapters 3 and 4, the problems of fault detection and that of designing checking sequences for sequential machines were discussed in detail. Given a sequential machine $M$, approaches are developed in this chapter that utilize the principle of machine modification through augmentation of extra inputs and outputs to considerably simplify the aforesaid problem of fault detection and that of designing checking sequences. The sequential machine $M'$ that results from augmentation of extra input and extra output logic to the original machine $M$ evidently includes $M$. One major advantage achieved through these approaches is that the original machine of which the faults are to be detected need not even be a strongly connected machine, because the modified machine is always strongly connected. Further, even if the original machine does not possess distinguishing sequences (for which the existing techniques based on preset experiments on the original machine are almost always very much involved and the designed checking sequences are too lengthy), the modified machine possesses distinguishing sequences, and thus simultaneously renders the problem of fault detection and of checking sequence design much easier. An added advantage of the present approach results from a coverage of the fault types not usually included in most checking experiments. In particular, in situations, where faults occurring in a sequential machine cause an increase in the total number of states of the machine, the methods of this chapter offer definite advantages. This latter type of machine faults may occur as we shall see later only in the case where a sequential machine has its number of states $q$ such that $2^{m-1} < q < 2^m$, thereby having some of the $m$-tuples of the $m$ state variables left unassigned.

## 5.2 Improved methods of fault detection in sequential machine using counter cycle

The first approach we shall present is based upon modification of the given sequential machine $M$ through addition of extra inputs and extra outputs. In this case, methods are developed for the fault diagnosis and for the design of checking sequences, taking into consideration the case where faults occurring in the machine may cause an increase in the number of states of the machine. Upper bounds on the lengths of the checking sequences are also established.

Let the sequential machine $M$ be defined by the quintuple $M=\langle I, S, O, f, g\rangle$ where $I$, $S$, and $O$ are, respectively, the input, state, and output alphabets of the machine, and $f$ and $g$ denote, respectively, the next state and output mappings. The machine considered here is a finite, deterministic, synchronous sequential machine of the *Mealy model*†, and is also assumed to be reduced and completely specified.

### 5.2.1 *Faults not resulting in an increase of states*

Assume that the total number of states of the machine $M$ is $q = 2^m$. In this case, each state of the machine is assigned exactly one of the distinct $m$-tuples of the $m$ state variables, with none of the $m$-tuples being left unused. The modification of the machine $M$ is carried out by augmenting only one extra input symbol $I_e$, and $\lceil \log_2 q \rceil$ output terminals. The $m$-tuples of the $m$ state variables are assigned to the states of $M$ as follows: Assign the $m$-tuple 00 . . . 001 to the state $S_1$, 00 . . . 010 to the state $S_2$, and so on upto the state $S_{q-1}$, while the all zero combination 00 . . . 000 is assigned to the $q$th state $S_q$. The next state and output mappings corresponding to the special input symbol $I_e$ are defined as given below.

$$f(S_i, I_e) = S_{i+1}; \quad g(S_i, I_e) = i;$$

$$f(S_q, I_e) = S_1; \quad g(S_q, I_e) = 0; \quad 1 \leqslant i \leqslant q-1$$

The outputs corresponding to the augmented input $I_e$ will be observed at the augmented output terminals, the decimal value of the output being $i$, an element of the set $\{0, 1, 2, \ldots, q-1\}$.

It is evident that the input symbol $I_e$ is a distinguishing sequence of length 1 for the modified machine $M'$, since the response for any state $S_i$ belonging to the state set $\{S_1, S_2, \ldots, S_q\}$ is a distinct decimal number $i$ assumed by the augmented output terminals. The following assumptions regarding the machine failures are relevant in this context.

(i) When faults occur in the given machine $M$, the machine simply operates as another sequential machine.

(ii) No fault can occur associated with the next state and output transitions corresponding to the special input symbol applied at the augmented input terminal.

(iii) During the fault-detection experiment, no new fault can occur.

The following steps of the procedure can conveniently be utilized in

†In a *Mealy model machine*, or simply *Mealy machine*, the next state is dependent on the present state and the present input, and the present output is also dependent on the present state and the present input.

the detection of faults, and finally in the design of checking sequences for the machine $M$.

*Step 1*: Apply the input symbol $I_e$ at the augmented input terminal repeatedly until an output of $O$ appears at the augmented output terminals. This operation ensures that the machine $M$ is transferred to the initial state $S_1$.

*Step 2*: Apply the input symbol $I_e$ at the augmented input terminal repeatedly, and record the corresponding output responses at the augmented output terminals until an output of $O$ appears at the augmented output terminals. This again ensures that the machine $M$ is brought back to the initial state $S_1$. The number of times the input $I_e$ has to be applied corresponds to the number of states of the machine.

*Step 3*: Start at the initial state $S_1$ of $M$ to which the machine is already set; apply the input symbol $I_i$ at the original input terminals and observe the corresponding output $O_i$ that appears at the original output terminals. This operation transfers the state $S_1$ to some other state $S_j$ of $M$. To check whether the next state transition of $S_1$ under input $I_i$ is nonfaulty, apply the special input symbol $I_e$ at the augmented input terminal, and note the corresponding output at the augmented output terminals. The application of $I_e$ at $S_j$ transfers the state $S_j$ to the state $S_{j+1}$, for $1 \leqslant j \leqslant q - 1$, and to the state $S_1$, for $S_j = S_q$. Finally, again apply the special input symbol $I_e$ the required number of times until the machine $M$ is brought back to the state $S_1$, as can be determined by checking the output at the augmented output terminals.

*Step 4*: Continue with Step 3 for every other state $S_k$, $k = 2, 3, \ldots, q$, of $M$, and for every state $S_j$, for all input symbols $I_j$, $j = 1, 2, \ldots, p$.

Based on a knowledge of these steps, the checking sequence for the machine $M$ can be designed quite readily. The entire checking sequence for the machine $M$ can be split into number of subsequences of which the first part will consist of only repeated occurrences of the special input symbol $I_e$. The maximum length of this subsequence, for a $q$-state machine $M$, is $2q$ (Steps 1 and 2). The remainder portion of the checking sequence is designed to check for every state of $M$, the next state transitions for all the original input symbols. For every state $S_i$ and every input $I_i$, this part of the checking sequence will consist of input symbol $I_i$, followed by special input symbol $I_e$, both applied only once, followed again by $I_e$, to bring the machine back to the state $S_i$. This part of the checking sequence, for every state and every input, has thus a maximum length of $(1 + 1 + q - 1)$ or $q + 1$ symbols. The subsequent portion of the checking sequence will consist of similarly designed subsequences, constructed for all the states and all the inputs of $M$. The input sequence being designed, the corresponding output sequence part of the checking sequence is obtained automatically. The mode of construction of the checking sequence simultaneously provides an upper bound on the length of such a sequence. For a $q$-state machine $M$ with $p$ input symbols, this upper bound can be easily found as $2q + (q + 1)pq$. The upper bound on the length of the checking sequence as obtained in [160] is $2q + (2q)pq$. Hence, the present method offers an improvement.

EXAMPLE: Let us consider a sequential machine $M_1$ as shown in Table 5.1. Its modified version $M_1'$ is shown in Table 5.2. The modification has been effected according to the scheme as discussed above.

Table 5.1 Machine $M_1$

| Present states | Next states and present outputs | |
|---|---|---|
| | $I_1 = 0$ | $I_2 = 1$ |
| $S_1$ | $S_2$, 0 | $S_4$, 0 |
| $S_2$ | $S_1$, 0 | $S_2$, 0 |
| $S_3$ | $S_4$, 1 | $S_1$, 0 |
| $S_4$ | $S_4$, 1 | $S_3$, 0 |

Table 5.2 Modified machine $M_1'$

| Present states | Next states and present outputs | | |
|---|---|---|---|
| | $I_1 = 0$ | $I_2 = 1$ | $I_3 = I_e$ |
| $S_1$(01) | $S_2$, 0 | $S_4$, 0 | $S_2$, 1 |
| $S_2$(10) | $S_1$, 0 | $S_2$, 0 | $S_3$, 2 |
| $S_3$(11) | $S_4$, 1 | $S_1$, 0 | $S_4$, 3 |
| $S_4$(00) | $S_4$, 1 | $S_3$, 0 | $S_1$, 0 |

For the machine $M_1$ as shown in Table 5.1, there is no distinguishing sequence. The designing of checking sequence for this machine according to the method of Kohavi and Lavallee [133], making the machine a DD machine, has already been discussed in Chapter 4. Here, the checking sequence for the same machine is designed according to the procedure outlined above. Suppose the machine was initially in state $S_1$; then Step 1 requires four applications of the special input symbol $I_e$. The entire checking sequence will be as follows:

| | | | | | | | | | |
|---|---|---|---|---|---|---|---|---|---|
| Input : | $I_e$ | $I_e$ | $I_e$ | $I_e$ | $I_e$ | $I_e$ | $I_e$ | $I_e$ | |
| State : | $S_1$ | $S_2$ | $S_3$ | $S_4$ | $S_1$ | $S_2$ | $S_3$ | $S_4$ | $S_1$ |
| Output: | (1) | (2) | (3) | (0) | (1) | (2) | (3) | (0) | |

Step 1 (first four outputs); Step 2 (last four outputs)

The third and fourth steps in combination will give the following transition checking part of the checking sequence.

| Input : | | 0 | $I_e$ | $I_e$ | $I_e$ | 1 | $I_e$ | $I_e$ | 0 | $I_e$ | 1 |
|---|---|---|---|---|---|---|---|---|---|---|---|
| State : | $S_1$ | $S_2$ | $S_3$ | $S_4$ | $S_1$ | $S_4$ | $S_1$ | $S_2$ | $S_1$ | $S_2$ | |
| Output: | | 0 | (2) | (3) | (0) | 0 | (0) | (1) | 0 | (1) | 0 |

(Contd.)

| Input : | | $I_e$ | 0 | $I_e$ | $I_e$ | $I_e$ | 1 | $I_e$ | $I_e$ | $I_e$ |
|---|---|---|---|---|---|---|---|---|---|---|
| State : | $S_2$ | $S_3$ | $S_4$ | $S_1$ | $S_2$ | $S_3$ | $S_1$ | $S_2$ | $S_3$ | |
| Output: | | (2) | 1 | (0) | (1) | (2) | 0 | (1) | (2) | (3) |

(Contd.)

| Input : | | 0 | $I_e$ | $I_e$ | $I_e$ | $I_e$ | 1 | $I_e$ |
|---|---|---|---|---|---|---|---|---|
| State : | $S_4$ | $S_4$ | $S_1$ | $S_2$ | $S_3$ | $S_4$ | $S_3$ | $S_4$ |
| Output: | | 1 | (0) | (1) | (2) | (3) | 0 | (3) |

The above scheme requires the application of 34 input symbols, which is less compared to the upper bound of 48 input symbols for such machines. Here the number within the parentheses represents the decimal value of the output that appears at the augmented output terminals.

### 5.2.2 *Faults resulting in an increase of states*

This part considers the problem of fault detection and that of designing checking sequences for a sequential machine, with the possibility of faults resulting in an increase of machine states. As has already been mentioned, in the present case the number of states $q$ of the machine satisfies the inequality $2^{m-1} < q < 2^m = q'$, $m$ being the number of memory elements. The required modification of the machine is carried out by adding two extra inputs $I_{e1}$, $I_{e2}$, the number of added output terminals being the same as before, i.e., $\lceil \log_2 q \rceil$. The next state and the output functions corresponding to the added inputs $I_{e1}$, $I_{e2}$ are:

$$f(S_i, I_{e1}) = S_{i+1}; \quad g(S_i, I_{e1}) = i;$$

$$f(S_q, I_{e1}) = S_1; \quad g(S_q, I_{e1}) = q;$$

$$f(S_r, I_{e1}) = S_r; \quad g(S_r, I_{e1}) = 0; \quad q+1 \leqslant r \leqslant q'; \quad 1 \leqslant i \leqslant q-1;$$

$$f(S_i, I_{e2}) = S_{i+1}; \quad g(S_i, I_{e2}) = i;$$

$$f(S_{q'}, I_{e2}) = S_1; \quad g(S_{q'}, I_{e2}) = 0; \quad 1 \leqslant i \leqslant q' - 1_0$$

The $m$-tuples of the state variables are assigned to the states of $M$ exactly as before, i.e., a state $S_i$ of $M$ is assigned an $m$-tuple of which the decimal value is $i$. The remaining unused $m$-tuples, $q' - q$, excepting the all zero combination, are each designated by $S_{q+1}, S_{q+2}, \ldots, S_{q'-1}$, where the number $j$ in each $S_j$ is again the decimal value of the $m$-tuple designated by $S_j$, while the all zero combination is designated by $S_{q'}$. The outputs corresponding to the augmented inputs will appear at the augmented output terminals.

Whenever the transition from a state $S_k$, $1 \leqslant k \leqslant q$, under an input $I_j$ is faulty, and is a state $S_m$, $q + 1 \leqslant m \leqslant q'$, the application of the special input symbol $I_{e1}$ at the augmented input terminal will produce an output of $O$ at the augmented output terminals. The exact state $S_m$ to which the transition has taken place can be definitely identified by next applying the second input $I_{e2}$ at the augmented input terminal, and noting the corresponding output at the augmented output terminals. Thus the detection of faults associated with the next state transitions offers no special problem. The checking sequence can also be constructed readily, since the principle of construction of the checking sequence is almost similar to the one discussed earlier.

Instead of using two different inputs, a somewhat different scheme based on the use of only a single input to detect faults including even those that may result in an increase of states of the machine is also practicable, and may even be simpler. In this case, the next state and output functions corresponding to this special input will be the same as that for the input $I_{e2}$ as discussed above. The $m$-tuples of the state variables are assigned to the states of $M'$ as before, i.e., the $m$-tuple 00...001 is assigned to the state $S_1$, 00...010 to the state $S_2$, and so on upto the state $S_q$, while the remaining unused tuples, $q' - q$, excepting the all zero combination, are each designated by $S_{q+1}, S_{q+2}, \ldots, S_{q'-1}$ where the number $j$ in each $S_j$ is the decimal value of the $m$-tuple designated by $S_j$, the all zero combination being designated by $S_{q'}$. The next state and output functions corresponding to the special input symbol $I_e$ are defined as follows:

$$f(S_i, I_e) = S_{i+1}; \quad g(S_i, I_e) = i;$$

$$f(S_{q'}, I_e) = S_1; \quad g(S_{q'}, I_e) = 0; \quad 1 \leqslant i \leqslant q' - 1.$$

The outputs corresponding to the augmented input $I_e$ appear at the augmented output terminals, the decimal value of the output being $\beta$, $\beta \in \{0, 1, \ldots, q' - 1\}$. The modified machine is evidently a CC machine. The input symbol $I_e$ happens to be distinguishing sequence of length 1 for the modified machine. Assume here that no faults can occur in the machine $M'$ associated with the next state transitions and outputs corresponding to the input $I_e$, and also during the fault detection experiment no new faults can occur. As an illustration, consider now the transition table of a machine $M_2$ as shown in Table 5.3. Table 5.4 gives the transition table of the modified machine $M_2'$.

A procedure for the detection of faults and in designing checking sequences based on the above scheme for a given sequential machine $M$ is now given below.

*Table 5.3 Machine $M_2$

| Present states | Next states and present outputs | |
|---|---|---|
| | $I_1 = 0$ | $I_2 = 1$ |
| $S_1(01)$ | $S_2$, 1 | $S_3$, 0 |
| $S_2(10)$ | $S_2$, 0 | $S_3$, 0 |
| $S_3(11)$ | $S_2$, 0 | $S_1$, 1 |

*Table 5.4 Machine $M_2$

| Present states | Next states and present outputs | | |
|---|---|---|---|
| | $I_1 = 0$ | $I_2 = 1$ | $I_3 = I_e$ |
| $S_1(01)$ | $S_2$, 1 | $S_3$, 0 | $S_2$, 1 |
| $S_2(10)$ | $S_2$, 0 | $S_3$, 0 | $S_3$, 2 |
| $S_3(11)$ | $S_2$, 0 | $S_1$, 1 | $S_4$, 3 |
| $S_4(00)$ | — | — | $S_1$, 0 |

*Step* (*a*): Apply the special input symbol $I_e$ at the augmented input terminal repeatedly until an output of $O$ appears at the augmented output terminals. This ensures that the machine $M$ is at the initial state $S_1$.

*Step* (*b*): Start at $S_1$. Apply $I_e$ at the augmented input terminal $k - 1$ times, $k \leqslant q$, and record the corresponding output responses produced at the augmented output terminals. This operation transfers $M$ from $S_1$ to a state $S_k$.

*Step* (*c*): At the state $S_k$, apply the input symbol $I_i$ at the original input terminals, and observe the corresponding output $O_i$ at the original output terminals. This transfers $M$ from $S_k$ to some other state $S_j$. Next, apply $I_e$ at the augmented input terminal once, and note the resulting output at the augmented output terminals. If the transition to $S_j$ was nonfaulty, then $M$ can be transferred upon the application of $I_e$ from $S_j$ to $S_{j+1}$, $j \leqslant q$. If the transition to $S_j$ was faulty, then upon the application of $I_e$, the machine $M$ is transferred from $S_j$ either to $S_{j+1}$, $1 \leqslant j \leqslant q' - 1$, or to $S_1$, for $j = q'$. The response produced at the augmented output terminals definitely identifies the state $S_j$.

*Step* (*d*): Continue (b) and (c) for every state $S_k$, $k = 1, 2, \ldots, q$, of $M$, and for every state $S_m$, for all inputs $I_i$, $i = 1, 2, \ldots, p$.

*Illustrations (Tables 5.3 and 5.4 etc.) in this section are from the paper referred [59] in the bibliography and are presented with the permission of IEE (copyright © 1978 IEE).

Based on the knowledge of these steps, the checking sequence for $M$ can also be designed quite readily. Let us now design a checking sequence for the machine $M_2$ as shown in Table 5.3 with the help of the augmented machine $M_2'$ shown in Table 5.4. The first step requires at most four consecutive applications of the special input symbol $I_e$, if the machine is in the state $S_1$ before the application of $I_e$, whereas it requires three consecutive applications of $I_e$ if the machine is in the state $S_2$, and so on. The following input-output sequence represents a checking sequence for the machine $M$ of Table 5.3, designed according to the scheme of fault detection mentioned above.

| | | | | | | | | | |
|---|---|---|---|---|---|---|---|---|---|
| Input : | $I_e$ | $I_e$ | $I_e$ | $I_e$ | $I_e$ | 0 | $I_e$ | 0 | |
| State : | $S_1$ | $S_2$ | $S_3$ | $S_4$ | $S_1$ | $S_2$ | $S_2$ | $S_3$ | |
| Output: | (1) | (2) | (3) | (0) | (1) | 0 | (2) | 0 | (Contd.) |
| Input : | $I_e$ | 1 | $I_e$ | 1 | $I_e$ | $I_e$ | 0 | $I_e$ | |
| State : | $S_2$ | $S_3$ | $S_1$ | $S_2$ | $S_3$ | $S_4$ | $S_1$ | $S_2$ | |
| Output: | (2) | 1 | (1) | 0 | (3) | (0) | 1 | (2) | (Contd.) |
| Input : | $I_e$ | $I_e$ | 1 | $I_e$ | | | | | |
| State : | $S_3$ | $S_4$ | $S_1$ | $S_3$ | $S_4$ | | | | |
| Output: | (3) | (0) | 0 | (3) | | | | | |

In the checking sequence outlined above, the special input symbol $I_e$ has been utilized as a distinguishing sequence of length one for the modified machine. Thus the advantages of utilizing $I_e$ as a homing sequence of length one can also be achieved simultaneously.

## 5.3 Fault detection in Moore model sequential machine utilizing counter cycle

In the approaches of Murakami et al. [160], the fault detection problem in *Moore model*† sequential machine was not considered. The applicability of counter cycle technique to the fault detection in Moore model sequential machine will be discussed here. The method will also take into consideration the types of faults which may result in an increase in the number of states of the machine.

A Mealy model is a pulse-input/pulse-output machine whereas a Moore model is a pulse-input/level-output machine; hence physically the two types are never identical. As the problem of fault detection involves only the observation of the input/output behaviour of the machine to determine whether or not the machine is operating correctly, one may seek

†In a *Moore model machine*, or simply *Moore machine*, the next state is dependent on the present state and the present input, but the present output is dependent only on the present state.

to design the checking sequence of a Moore model by considering the state table of an equivalent Mealy model. But the checking sequence in practice may not result in an isomorphic table of the equivalent Mealy model because of the faults present in the machine itself. In this context, it may be recalled that due to the state splitting in the conversion of a Mealy model to an equivalent Moore model the number of states are increased, though in certain special cases the number may remain the same. Therefore it becomes evident that the flow table of an equivalent Moore model which follows from that of the Mealy model obtained from the observed input/output sequence in presence of faults in the machine, will not be the actual transition table isomorphic to the original Moore model machine. Thus for designing the checking sequence for a Moore model machine an alternative approach is suggested. The approach is to design a new CC machine which contains the original Moore model by augmenting one input and one output terminal to the original machine so that with respect to the special input applied to the augmented input terminal and the output observed at the augmented output terminal, the behaviour of the augmented machine will be like that of a Mealy model. It is to be noted that in the modified machine the Moore model behaviour of the original machine will be preserved with respect to the original input and output terminals of the machine.

### 5.3.1 *Faults not increasing the number of states*

This is possible when the number of states of the machine $M$ is $q = 2^m$, with none of the $m$-tuples of the $m$ state variables being left unused, and each state of the machine is assigned exactly one of the distinct $2^m$ $m$-tuples. The modification is effected by augmenting one input terminal for the special input $I_e$, and only one output terminal. The outputs corresponding to the original input symbols will appear in the original output terminals only, while the outputs corresponding to the special input $I_e$ will appear in the augmented output terminal only.

The next state and the output functions corresponding to $I_e$ is as follows:

$$f(S_i, I_e) = S_{i+1}; \quad g(S_i, I_e) = 0; \quad 1 \leqslant i \leqslant q-1;$$
$$f(S_q, I_e) = S_1; \quad g(S_q, I_e) = 1$$

Let us assume that (i) when faults occur in a given machine, it simply operates as another sequential machine, (ii) no faults occur associated with the next state transitions and outputs corresponding to the special input symbol $I_e$, and (iii) during the fault detection experiment no new faults can occur. The undernoted steps of the procedure will be found satisfactory when the original machine happens to be a Moore model machine.

*Step 1*: Apply the input $I_e$ in the augmented input terminal repeatedly until an output of 1 appears in the augmented output terminal.

*Step 2*: Again apply the input $I_e$ in the augmented input terminal repeatedly until an output of 1 appears at the augmented output terminal. The number of times $I_e$ has to be applied equals the number of states of the machine.

*Step 3*: To check the transitions from any state $S_i$, $S_i \in \{S_1, S_2, \ldots, S_q\}$, the following subsequences are applied.

(a) The corresponding $T_i$ sequence is applied to set the machine to the state $S_i$. {It can be recalled that for a CC machine with a counter cycle in input $I_e$, $T_i$ sequence of the state $S_i$ is an input-output sequence such that (i) the input sequence consists of $(i - 1)$ consecutive symbols $I_e$, and (ii) the output sequence consists of $(i - 1)$ consecutive symbols $O$ [160].}

(b) An input $I_j$ is applied at the original input terminals and the output $O_j$ is observed at the original output terminals. $O_j$ is the present output associated with the $I_j$th successor of the state $S_i$, i.e., with the state $S_x$, say; thus $I_j - O_j$ becomes the respective input-output pair.

(c) The state $S_x$ can be identified by applying the special input symbol $I_e$ repeatedly until an output of 1 appears at the augmented output terminal. If 1 is obtained in this process at the $k$th application of $I_e$, then the state $S_x$ is identified as $S_{q-k+1}$.

*Step 4*: All the transitions from all $S_i \in \{S_1, S_2, \ldots, S_q\}$ under all inputs $I_j \in \{I_1, I_2, \ldots, I_p\}$ are checked similarly by repeating the above Step 3 of the procedure.

As can be seen, the procedure is an extension of the works of Murakami et al. [160] to the case of Moore model sequential machines. The upper bound on the length of the checking sequence thus remains the same as that obtained by Murakami et al. [160].

EXAMPLE: Consider a 4-state Moore model sequential machine $M_3$ as shown in Table 5.5. The modified machine $M_3'$ is shown in Table 5.6.

*Table 5.5 Machine $M_3$

| Present states | Next states | | Present outputs |
|---|---|---|---|
| | $I_1 = 0$ | $I_2 = 0$ | |
| $S_1$ | $S_2$ | $S_3$ | 0 |
| $S_2$ | $S_4$ | $S_1$ | 0 |
| $S_3$ | $S_3$ | $S_2$ | 1 |
| $S_4$ | $S_1$ | $S_2$ | 0 |

*Illustrations (Tables 5.5, 5.6 etc.) in this section are from the paper referred [23] in the bibliography and are presented with the permission of IEE (copyright © 1979 IEE).

*Table 5.6 Machine $M_3'$

| Original Moore machine | | | | | Augmented Mealy machine | |
|---|---|---|---|---|---|---|
| Present states | Next states | | Present outputs | | Present states | Next states and the outputs corresponding to the special input $I_e$ |
| | $I_1 = 0$ | $I_2 = 1$ | | | | |
| $S_1$ | $S_2$ | $S_3$ | 0 | + | $S_1$ | $S_2$, 0 |
| $S_2$ | $S_4$ | $S_1$ | 0 | | $S_2$ | $S_3$, 0 |
| $S_3$ | $S_3$ | $S_2$ | 1 | | $S_3$ | $S_4$, 0 |
| $S_4$ | $S_1$ | $S_2$ | 0 | | $S_4$ | $S_1$, 1 |

If the checking sequence is designed for the machine $M_3$ shown in Table 5.5 with the aid of its modified version $M_3'$ shown in Table 5.6. Step 1 needs at most the applications of four repeated special input symbol $I_e$. The checking sequence is shown below.

Input : $I_e$ $I_e$ $I_e$ $I_e$ $I_e$ $I_e$ $I_e$ $I_e$

State : $S_1$ $S_2$ $S_3$ $S_4$ $S_1$ $S_2$ $S_3$ $S_4$

Output: 0 0 0 1 0 0 0 1 (Contd.)

Input : 0 $I_e$ $I_e$ $I_e$ 1 $I_e$ $I_e$ $I_e$ 0 $I_e$

State : $S_1$ $S_2$ $S_3$ $S_4$ $S_1$ $S_3$ $S_4$ $S_1$ $S_2$ $S_4$

Output: 0 0 0 1 1 0 1 0 0 1 (Contd.)

Input : $I_e$ 1 $I_e$ $I_e$ $I_e$ $I_e$ $I_e$ $I_e$ 0

State : $S_1$ $S_2$ $S_1$ $S_2$ $S_3$ $S_4$ $S_1$ $S_2$ $S_3$ $S_3$

Output: 0 0 0 0 0 1 0 0 1 (Contd.)

Input : $I_e$ $I_e$ $I_e$ $I_e$ 1 $I_e$ $I_e$ $I_e$

State : $S_4$ $S_1$ $S_2$ $S_3$ $S_2$ $S_3$ $S_4$

Output: 0 1 0 0 0 0 0 1 (Contd.)

Input : $I_e$ $I_e$ $I_e$ 0 $I_e$ $I_e$ $I_e$ $I_e$ $I_e$

State : $S_1$ $S_2$ $S_3$ $S_4$ $S_1$ $S_2$ $S_3$ $S_4$ $S_1$

Output: 0 0 0 0 0 0 0 1 0 (Contd.)

Input : $I_e$ $I_e$ 1 $I_e$ $I_e$ $I_e$

State : $S_2$ $S_3$ $S_4$ $S_2$ $S_3$ $S_4$ $S_1$

Output: 0 0 0 0 0 1

Note: Since in the Moore machine the output depends on state only, arrow indications are placed to indicate these outputs corresponding to the states and observed at the original output terminal of the machine.

### 5.3.2 *Faults increasing the number of machine states*

This may be possible if the number of states of the sequential machine satisfies the inequality $2^{m-1} < q < 2^m = q'$, with some of the $m$-tuples of the $m$ state variables being left unassigned. In this case the modification of the machine is effected by augmenting two extra inputs $I_{e1}$, and $I_{e2}$, and only one output variable. It is possible to apply the special inputs $I_{e1}$, and $I_{e2}$ to a single augmented input terminal. Here also the outputs of the two special input symbols will be associated with the augmented output terminal only, while the outputs corresponding to the original input symbols will be associated with the original output terminals only.

The next state and the output mappings of the two special input symbols are as follows:

$$f(S_i, I_{e1}) = S_{i+1}; \quad g(S_i, I_{e1}) = 0; \quad 1 \leqslant i \leqslant q-1;$$

$$f(S_q, I_{e1}) = S_1; \quad g(S_q, I_{e1}) = 1;$$

$$f(S_r, I_{e1}) = S_r; \quad g(S_r, I_{e1}) = 0; \quad q+1 \leqslant r \leqslant q';$$

$$f(S_i, I_{e2}) = S_{i+1}; \quad g(S_i, I_{e2}) = 0; \quad 1 \leqslant i \leqslant q'-1;$$

$$f(S_{q'}, I_{e2}) = S_1; \quad g(S_{q'}, I_{e2}) = 1$$

As can be seen from the above, the checking sequence of the machine $M$ can be designed with the help of the counter cycle in $I_{e1}$. Whenever the next state transitions from any state $S_k$, $1 \leqslant k \leqslant q$, under an input symbol $I_j$, $1 \leqslant j \leqslant p$, is faulty and is a state $S_r$, $q+1 \leqslant r \leqslant q'$, then at $S_r$, the repeated applications of $I_{e1}$ will yield 0's only and thus $S_r$ will not be identified by the applications of $I_{e1}$ only. But the repeated applications of $I_{e2}$ at the augmented input terminal until an output of 1 appears at the augmented output terminal will definitely identify the state $S_r$. If in the above process, at the $n$th step of applications of $I_{e2}$, a 1 appears at the augmented output terminal, then $S_r = S_{q'-n+1}$.

## †5.4 Readily diagnosable sequential machines with extra inputs and extra outputs

Given a sequential machine $M$, the approaches are developed in this section, that utilize the principle of machine modification to considerably simplify the problem of fault detection and that of designing checking sequences. In the first half of this section we shall consider the case where the total number of states of the machine $M$ becomes exactly equal to $2^m$, though a violation of this condition in no way invalidates the generality of the results obtained, so long as our assumption that the faults do not cause an increase of machine states remains. A well defined procedure for

†Illustrations (Tables 5.7, 5.8, 5.9, 5.10 etc.) in Section 5.4 are from the paper referred [60] in the bibliography and are reprinted with permission from Pergamon Journals Ltd. (copyright © 1979 Pergamon Press).

designing checking sequences for such types of sequential machines is developed. The second part of the section deals with the case that involves fault detection for those sequential machines for which the number of states $q$ exceeds $2^{m-1}$, but does not equal $2^m$, with the possibility of faults resulting in an increase of the number of states of the machines. A systematic procedure for fault diagnosis for this type of machines is also developed [60].

### 5.4.1 *Faults not resulting in an increase of machine states*

Assume that the total number of states of the machine $M$ is $q = 2^m$. In this case, each state of the machine is assigned exactly one of the $2^m$ distinct $m$-tuples of the $m$ state variables, with none of the $m$-tuples being left unused. We propose to modify the given machine $M$ by augmenting two extra input symbols $I_{e1}$ and $I_{e2}$, and $\lceil \log_2 q \rceil$ output terminals, where $\lceil r \rceil$ denotes the smallest integer greater than or equal to $r$ (which in this particular case equals $m$). The $m$-tuples of the state variables are arbitrarily assigned to the states of $M$ as follows. Assign the $m$-tuple 00 . . . 001 to the state $S_1$; 00 . . . 010 to the state $S_2$; and so on upto the state $S_{q-1}$, while the all zero combination 00 . . . 000 is assigned to the $q$th state $S_q$. The next state and output mappings corresponding to the special input symbols $I_{e1}$, $I_{e2}$ are defined as given below.

$$f(S_i, I_{e1}) = S_{i+1}; \quad g(S_i, I_{e1}) = i;$$

$$f(S_q, I_{e1}) = S_1; \quad g(S_q, I_{e1}) = 0;$$

$$f(S_i, I_{e2}) = S_1; \quad g(S_i, I_{e2}) = i;$$

$$f(S_q, I_{e2}) = S_1; \quad g(S_q, I_{e2}) = 0; \quad 1 \leqslant i \leqslant q-1$$

The outputs corresponding to the augmented inputs $I_{e1}$, $I_{e2}$ will appear at the augmented output terminals, the decimal value of the output being $\alpha$, an element of the set $\{0, 1, \ldots, q-1\}$. Here it becomes evident that in the modified machine $M'$ the following holds good, viz., (i) The input symbol $I_{e1}$ (as well as $I_{e2}$) is a distinguishing sequence of length 1 for the modified machine $M'$; (ii) The input symbol $I_{e2}$ is a synchronizing sequence of length 1 for the modified machine $M'$; (iii) Upon the application of an input sequence consisting of $q$ consecutive occurrences of the special input symbol $I_{e1}$, if a sequential machine $M'$ produces an output sequence at the augmented output terminals which is a concatenation of $q$ symbols $i_k$, $k = 1, 2, \ldots, q$, where each $i_k$ is distinct and an element of the set $\{0, 1, \ldots, q-1\}$, then the machine $M'$ must have at least $q$ internal states.

EXAMPLE: Consider a sequential machine $M_1$ as given by the transition table of Table 5.7. Table 5.8 gives the transition table of the modified machine $M_1'$. Note that the machine $M_1$ is reduced, but not strongly connected; the modified machine $M_1'$ is strongly connected since $M_1'$ is a cc machine for the input $I_{e1}$.

Table 5.7 Machine $M_1$

| Present states | Next states and present outputs | |
|---|---|---|
| | $I_1 = 0$ | $I_2 = 1$ |
| $S_1(01)$ | $S_3$, 0 | $S_2$, 0 |
| $S_2(10)$ | $S_2$, 1 | $S_3$, 0 |
| $S_3(11)$ | $S_2$, 1 | $S_1$, 1 |
| $S_4(00)$ | $S_4$, 0 | $S_3$, 1 |

Table 5.8 Machine $M_1'$

| Present states | Next states and present outputs | | | |
|---|---|---|---|---|
| | $I_1 = 0$ | $I_2 = 1$ | $I_{e1}$ | $I_{e2}$ |
| $S_1(01)$ | $S_3$, 0 | $S_2$, 0 | $S_2$, 1 | $S_1$, 1 |
| $S_2(10)$ | $S_2$, 1 | $S_3$, 0 | $S_3$, 2 | $S_1$, 2 |
| $S_3(11)$ | $S_2$, 1 | $S_1$, 1 | $S_4$, 3 | $S_1$, 3 |
| $S_4(00)$ | $S_4$, 0 | $S_3$, 1 | $S_1$, 0 | $S_1$, 0 |

The undernoted steps of the procedure can be conveniently utilized in the detection of faults and finally in the design of checking sequences for a given sequential machine $M$. Of course, certain assumptions are required to be made in this connection. The assumptions regarding the nature of faults in the machine as stated earlier in Section 5.2.1 of this chapter are considered to be relevant under the present context also.

*Step 1*: Apply the special input symbol $I_{e2}$ at the augmented input terminal only once and note the corresponding output symbol produced at the augmented output terminals. This operation ensures that the machine $M$ is transferred to the initial state $S_1$.

*Step 2*: Apply the special input symbol $I_{e1}$ at the augmented input terminal repeatedly, and record the corresponding output responses at the augmented output terminals until an output of $O$ appears at the augmented output terminals. This again ensures that the machine $M$ is brought back to the initial state $S_1$ after traversing through all its states. The number of times $I_{e1}$ has to be applied corresponds to the number of states $q$ of the machine.

*Step 3*: Start at the initial state $S_1$ of $M$ to which $M$ is already set; apply the special input symbol $I_{e1}$ at the augmented input terminal $k - 1$ times and record the corresponding output responses produced at the augmented output terminals. This operation transfers the machine $M$ from the state $S_1$ to the state $S_k$.

*Step 4*: At the state $S_k$ of $M$, apply the input symbol $I_i$ at the original input terminals, and observe the corresponding output $O_i$ that appears at the original output terminals. This operation transfers the machine $M$ from the state $S_k$ to some other state, say, $S_j$. To ensure that the next state transitions of $M$ from $S_k$ under input $I_i$ is nonfaulty, apply the special input symbol $I_{e2}$ at the augmented input terminal only once, and note the corresponding output at the augmented output terminals. This application of the special input symbol $I_{e2}$ at the state $S_j$ transfers $M$ from the state $S_j$ back to the state $S_1$. The response produced at the augmented output terminals definitely identifies the state $S_j$.

*Step 5*: Continue with Steps 3 and 4 for every state $S_k$, $k = 1, 2, \ldots, q$, of $M$ and for every state $S_m$, for all input symbols $I_i$, $i = 1, 2, \ldots, p$.

Based on a knowledge of these steps, the checking sequence for the machine $M$ can be designed quite readily. The entire checking sequence for $M$ can be split into a number of subsequences of which the first part will consist of the special input symbol $I_{e2}$ followed by repeated occurrences of the special input symbol $I_{e1}$. The total length of this subsequence, for a $q$-state machine $M$, is $q + 1$ (Steps 1 and 2). The remainder portion of the checking sequence is designed to check for every state of $M$, the next state transitions for all the original input symbols. For every state $S_i$ and for every input $I_i$, this part of the checking sequence will consist of $i - 1$ repeated occurrences of the special input symbol $I_{e1}$, followed by a single occurrence of the input symbol $I_i$ as well as of the special input symbol $I_{e2}$, for bringing the machine $M$ back to the initial state $S_1$. This part of the checking sequence has, for every state $S_i$ and every input $I_i$ of $M$, a length of $i + 1$ (Steps 3 and 4). The subsequent portion of the checking sequence will consist of similarly designed subsequences, constructed for all the states and all of the inputs of $M$. The input sequence being designed, the corresponding output sequence part of the checking sequence is obtained automatically. The mode of construction of the checking sequence simultaneously provides a bound on the length of such a sequence. For a $q$-state machine $M$ with $p$ input symbols, the total length $L$ of the checking sequence can be easily found as:

$$
\begin{aligned}
L &= (q + 1) + \left[\sum_{i=1}^{i=q} (i + 1)\right] p \text{ symbols} \\
&= (q + 1) + \left[\frac{(q + 1)(q + 2)}{2} - 1\right] p \text{ symbols} \\
&= (q + 1) + \tfrac{1}{2} p(q^2 + 3q) \text{ symbols.}
\end{aligned}
$$

EXAMPLE: According to the above procedure, one can design the checking sequence of the machine $M_1$ shown in Table 5.7 with the help of its augmented version $M_1'$ shown in Table 5.8. This checking sequence is shown below where the presence of a dash (—) in the state and output sequence represents a don't care condition.

Input : $I_{e2}$ $I_{e1}$ $I_{e1}$ $I_{e1}$ $I_{e1}$ 0 $I_{e2}$ 1

State : — $S_1$ $S_2$ $S_3$ $S_4$ $S_1$ $S_3$ $S_1$ $S_2$

Output: — (1) (2) (3) (0) 0 (3) 0 (Contd.)

$\longleftarrow$ Steps 1 & 2 $\longrightarrow$ $\leftarrow$Steps 3, 4 & 5—

Input : $I_{e2}$ $I_{e1}$ 0 $I_{e2}$ $I_{e1}$ 1 $I_{e2}$ $I_{e1}$ $I_{e1}$

State : $S_2$ $S_1$ $S_2$ $S_2$ $S_1$ $S_2$ $S_3$ $S_1$ $S_2$ $S_3$

Output: (2) (1) 1 (2) (1) 0 (3) (1) (2)

———— Steps 3, 4 & 5 ———— (Contd.)

Input : 0 $I_{e2}$ $I_{e1}$ $I_{e1}$ 1 $I_{e2}$ $I_{e1}$ $I_{e1}$ $I_{e1}$

State : $S_3$ $S_2$ $S_1$ $S_2$ $S_3$ $S_1$ $S_1$ $S_2$ $S_3$ $S_4$

Output: 1 (2) (1) (2) 1 (1) (1) (2) (3)

———— Steps 3, 4 & 5 ———— (Contd.)

Input : 0 $I_{e2}$ $I_{e1}$ $I_{e1}$ $I_{e1}$ 1 $I_{e2}$

State : $S_4$ $S_4$ $S_1$ $S_2$ $S_3$ $S_4$ $S_3$ $S_1$

Output: 0 (0) (1) (2) (3) 1 (3)

———— Steps 3, 4 & 5 ————$\rightarrow$

Note that in the above checking sequence, the output sequence part represents the decimal value of the output appearing at the augmented output terminals when shown inside the bracket. The above checking sequence requires the application of 33 input symbols.

### 5.4.2 *Faults resulting in an increase of machine states*

In this part, we consider the problem of fault detection and that of designing checking sequences for a sequential machine, taking into consideration the possibility of faults resulting in an increase of machine states. As is already mentioned, for a machine realized by binary devices, the possibility of occurrence of failures that may result in an increase of states is indeed rather small when the total number of machine states $q$ is an integral power of 2, since in this case physical creation of new state variables would have been implied. However, if the number of machine states does not happen to be an integral power of 2, or more than $\lceil \log_2 q \rceil$ state variables are used in the realization, such failures are likely to occur. Without any loss of generality, assume that the number of states $q$ of the machine $M$ satisfies the condition $2^{m-1} < q < 2^m = q'$, $m$ being the number of memory elements used in the realization. The required modification for the machine $M$ is done by adding two extra inputs $I_{e1}$, $I_{e2}$, the number of added output terminals being the same as before, i.e., $\lceil \log_2 q \rceil = m$. The next state and output functions corresponding to the added inputs $I_{e1}$, $I_{e2}$ are:

$$f(S_i, I_{e1}) = S_{i+1}; \quad g(S_i, I_{e1}) = i;$$
$$f(S_q, I_{e1}) = S_1; \quad g(S_q, I_{e1}) = q;$$
$$f(S_r, I_{e1}) = S_r; \quad g(S_r, I_{e1}) = 0;$$
$$f(S_j, I_{e2}) = S_1; \quad g(S_j, I_{e2}) = j;$$
$$f(S_{q'}, I_{e2}) = S_1; \quad g(S_{q'}, I_{e2}) = 0;$$
$$1 \leqslant i \leqslant q-1; \quad q+1 \leqslant r \leqslant q'; \quad 1 \leqslant j \leqslant q'-1$$

The $m$-tuples of the state variables are assigned to the states of $M'$ exactly as before, i.e., a state $S_i$ of $M$ is assigned an $m$-tuple of which the decimal value is $i$. The remaining unused $m$-tuples, $q' - q$, excepting the all zero combination, are each designated by $S_{q+1}, S_{q+2}, \ldots, S_{q'-1}$, where the number $j$ in each $S_j$ is again the decimal value of the $m$-tuple designated by $S_j$, while the all zero combination is designated by $S_{q'}$. The outputs corresponding to the augmented inputs appear at the augmented output terminals.

EXAMPLE: Table 5.9 shows the transition table for a sequential machine $M_2$ for which the number of states $q$ is not an integral power of 2. Table 5.10 gives the transition table of the modified machine $M_2'$.

Table 5.9 Machine $M_2$

| Present states | Next states and present outputs | |
|---|---|---|
| | $I_1 = 0$ | $I_2 = 1$ |
| $S_1(01)$ | $S_2$, 1 | $S_3$, 0 |
| $S_2(10)$ | $S_2$, 0 | $S_3$, 0 |
| $S_3(11)$ | $S_2$, 0 | $S_1$, 1 |

Table 5.10 Modified machine $M_2'$

| Present states | Next states and present outputs | | | |
|---|---|---|---|---|
| | $I_1 = 0$ | $I_2 = 1$ | $I_{e1}$ | $I_{e2}$ |
| $S_1(01)$ | $S_2$, 1 | $S_3$, 0 | $S_2$, 1 | $S_1$, 1 |
| $S_2(10)$ | $S_2$, 0 | $S_3$, 0 | $S_3$, 2 | $S_1$, 2 |
| $S_3(11)$ | $S_2$, 0 | $S_1$, 1 | $S_1$, 3 | $S_1$, 3 |
| $S_4(00)$ | — | — | $S_4$, 0 | $S_1$, 0 |

Whenever the transition from a state $S_k$, $1 \leqslant k \leqslant q$, under an input $I_i$ is faulty and is a state $S_m$, $q+1 \leqslant m \leqslant q'$ the application of the special input symbol $I_{e1}$ at the augmented input terminal will produce an output of 0 at the augmented output terminals. The exact state $S_m$ to which the transition has taken place can be definitely identified by next applying the second input $I_{e2}$ at the augmented input terminal, and noting the corresponding output at the augmented output terminals. Thus the detection of faults associated with the next state transitions offers no special problem. The checking sequence can also be constructed readily, since the principle of construction of the checking sequence is almost similar to that as has already been discussed previously.

The method as presented above provides a short, simple, and efficient checking experiment for sequential machines by adding two extra inputs and $\lceil \log_2 q \rceil$ output terminals. The work is evidently based upon the original works of Kohavi and Lavallee [133] and of Marakami et al. [160]. It must be emphasized here that in the process the compromise that is adopted, viz., added logic against lengthy experiments does not appear too costly and hence unacceptable. The amount of hardware required in introducing the extra input and output logic may be roughly estimated as follows: (i) no extra memory elements will be required; (ii) some additional combinational logic for realizing the state transitions and outputs associated with the extra inputs will be needed; and (iii) only one extra input and $\lceil \log_2 q \rceil$ output terminals need be required. The hard-ware requirements necessitated by (ii) and (iii) may be reduced to a certain extent by appropriate input and output assignments.

The length of the checking experiment as obtained through the approach of this section is considerably reduced. A comparison of the bound obtained by other workers [133, 160, 116, 85] reveals that the bound on the length of the experiment as obtained herein is better than all the others for small $p$ and $q$ (say, $p = 2$; $q = 8$), while for larger $p$, $q$ (say, $p = 50$; $q = 100$) the method of Fujiwara et al. [85] gives the best bound. This increase in length for large $p$, $q$ is due mostly to the larger length of the transfer sequence in the present method, of which the maximum length is only $\lceil \log_2 q \rceil$ in [85] and is $q - 1$ in our case, which may be considerable for large $q$. A substantial reduction in length of the checking experiment may be achieved by using an augmentation scheme that combines the better aspects of the present method with that of the method presented in [85] and discussed earlier in Section 4.6. For instance, by modifying a given machine with four extra inputs and $\lceil \log_2 q \rceil$ output terminals, of which the two inputs are the same as used here and the other two are as in [85] will reduce not only the lengths of the distinguishing sequence and synchronizing sequence, but also the length of the transfer sequence which contributes mostly to the overall length of the checking experiments.

## †5.5 Very easily testable/diagnosable sequential machine

We shall present in this section, a new scheme for modifying a given sequential machine $M$ with the help of four extra input symbols $I_{e1}$, $I_{e2}$, $I_{e3}$ and $I_{e4}$ and $\lceil \log_2 q \rceil$ output terminals. It can be recalled from Section 4.6 of Chapter 4 that recently Fujiwara et al. presented a modification scheme in a given sequential machine that utilizes two extra input symbols, and showed that a $q$-state machine $M$ after such modification possesses a distinguishing-cum-synchronizing sequence of length $\lceil \log_2 q \rceil$ symbols and a transfer sequence of length at most $\lceil \log_2 q \rceil$ symbols to take the machine from some state $S_1$ to a state $S_i$ for all $i$. Such augmentation makes the machine strongly connected, and gives a reduced bound on the length of the checking sequence which is approximately equal to $pq \lceil \log_2 q \rceil$ symbols for a $q$-state machine with $p$ input symbols. The machine was named an 'easily testable machine'. Given a sequential machine, an approach is developed in this section that considerably simplify the authors' procedure [60] to obtain a better checking sequence for the modified machine. In particular, the present approach remains applicable even in situations where faults occurring in a given machine may cause an increase in the number of states of the machine. A reduced bound on the length of the checking sequence is obtained by reducing the length of the synchronizing-cum-distinguishing sequence to only a single input symbol compared to $\lceil \log_2 q \rceil$ symbols as utilized in the method of Fujiwara et al. [85].

Let us take the case of a sequential machine where the number of states $q$ of the machine $M$ can be an integral power of 2, but without any loss of generality, assume that the number of states $q$ of the machine satisfies the inequality $2^{m-1} < q < 2^m = q'$, $m$ being the number of memory elements used in the realization.

The next state and output functions corresponding to the added inputs $I_{e1}$, and $I_{e2}$ are defined as follows:

$$f(S_i, I_{e1}) = S_1; \quad g(S_i; I_{e1}) = i; \quad 1 \leqslant i \leqslant q;$$
$$f(S_r, I_{e1}) = S_r; \quad g(S_r, I_{e1}) = 0; \quad q+1 \leqslant r \leqslant q';$$
$$f(S_j, I_{e2}) = S_1; \quad g(S_j, I_{e2}) = j; \quad 1 \leqslant j \leqslant q'-1;$$
$$f(S_{q'}, I_{e2}) = S_1; \quad g(S_{q'}, I_{e2}) = 0$$

The outputs corresponding to the augmented inputs $I_{e1}$ and $I_{e2}$ will appear at the augmented output terminals, the decimal value of the output being $\alpha$, $\alpha \in \{0, 1, 2, \ldots, q\}$ and $\beta$, $\beta \in \{0, 1, \ldots, q'-1\}$ for $I_{e1}$ and $I_{e2}$, respectively. The extra inputs $I_{e1}$, and $I_{e2}$ of the modified machine thus become distinguishing-cum-synchronizing sequence of length 1 for the original and modified machines, respectively.

The next state and output functions corresponding to the extra inputs

†Illustrations (Tables 5.11, 5.12 etc.) in Section 5.5 are from the paper referred [22] in the bibliography and are presented with the permission of IEE (copyright © 1979 IEE).

$I_{e3}$ and $I_{e4}$ will serve a purpose similar to that of the augmented input symbols in Section 4.6, and will thus provide a transfer sequence of length at most $\lceil \log_2 q \rceil$ symbols for the modified machine. These functions are defined as follows:

For each state $S_i$, with the state assignment $Y_1Y_2 \ldots Y_m$,

$$f(S_i, I_{e3}) = S_j; \quad f(S_i, I_{e4}) = S_k;$$

$$g(S_i, I_{e3}) = g(S_i, I_{e4}) = O_1; \quad \text{if } Y_m = 0;$$

$$= O_2; \text{ if } Y_m = 1;$$

where $O_1$ and $O_2$ may be 0 and 1 respectively, and the states $S_j$ and $S_k$ have the assignments $0Y_1Y_2 \ldots Y_{m-1}$ and $1Y_1Y_2 \ldots Y_{m-1}$ respectively.

As an illustration, consider the transition table of the machine $M_2$ as shown in Table 5.11. Table 5.12 gives the transition table of the modified machine $M_2'$. To understand the superiority of this method over that of

Table 5.11 Machine $M_2$

| Present states | Next states and present outputs | |
|---|---|---|
| | $I_1 = 0$ | $I_2 = 1$ |
| $(00)S_1$ | $S_2$, 1 | $S_1$, 1 |
| $(01)S_2$ | — | $S_3$, 0 |
| $(10)S_3$ | $S_2$, 0 | —, 1 |

Table 5.12 Machine $M_2'$

| Present states | Next states and present outputs | | | | | |
|---|---|---|---|---|---|---|
| | $I_1 = 0$ | $I_2 = 1$ | $I_{e1}$ | $I_{e2}$ | $I_{e3}$ | $I_{e4}$ |
| $(00)S_1$ | $S_2$, 1 | $S_1$, 1 | $S_1$, 1 | $S_1$, 1 | $S_1$, 0 | $S_3$, 0 |
| $(01)S_2$ | — | $S_3$, 0 | $S_1$, 2 | $S_1$, 2 | $S_1$, 1 | $S_3$, 1 |
| $(10)S_3$ | $S_2$, 0 | —, 1 | $S_1$, 3 | $S_1$, 3 | $S_2$, 0 | $S_4$, 0 |
| $(11)S_4$ | — | — | $S_4$, 0 | $S_1$, 0 | $S_2$, 1 | $S_4$, 1 |

Fujiwara et al. [85], the transition table of the machine we have used in [22] is taken to be isomorphic to that of Fujiwara et al. [85].

If the augmented machine $M'$ is predesigned according to the above-mentioned principle, then the checking experiment of the original machine $M$ can be designed very easily. For this purpose of designing checking experiment the special input symbol $I_{e1}$ can be used as a synchronizing-cum-distinguishing sequence of length 1. Further in this approach, the failures resulting in an increase in the number of states of the original machine $M$ can be properly diagnosed. Suppose, the application of a particular input symbol $I_j$ upon the state $S_j$ of the original machine $M$

causes a state transition which leads the machine to a faulty state $S_x$, such that $q+1 \leqslant x \leqslant q'$. Then at this stage the subsequent application of $I_{e1}$ will yield 0 for $S_x$. The same state $S_x$ can definitely be identified by applying $I_{e2}$ once, and noting the decimal value of the output in the augmented output terminals. In this way the machine can also be brought back to the initial state $S_1$ from the faulty state $S_x$.

When a sequential machine $M$ is considered to be embedded inside its modified version $M'$, generally it was discussed earlier how to design a checking sequence of the original machine $M$ which receives some important attributes from its modified version $M'$. It will be appropriate here to consider, like Fujiwara et al. [85], the designing aspect of the checking sequence for the entire modified machine $M'$. It becomes at once clear that the special input symbol $I_{e2}$ can serve the purpose of a distinguishing-cum-synchronizing sequence of length 1 for the modified machine $M'$. As an illustration let us now consider the designing of the checking sequence for the modified machine $M_2'$ of Table 5.12. The transfer sequences which will be required for this purpose are shown in Table 5.13, and the checking sequence of the machine $M_2'$ as constructed is shown next.

Table 5.13 Transfer or $T(i)$ sequence for the Modified Machine $M_2$

| $T(1)$ | $T(2)$ | $T(3)$ | $T(4)$ |
|---|---|---|---|
| $\lambda$ (null) | $I_{e4}\ I_{e3}$ | $I_{e4}$ | $I_{e4}\ I_{e4}$ |

Checking sequence for the modified machine $M_2'$ is shown below:

Input : $I_{e2}$ $T(1)$ $I_{e2}$ $I_{e2}$ $T(2)$ $I_{e2}$ $I_{e2}$
State : — $S_1$ $S_1$ $S_1$ $S_1$ $S_2$ $S_1$ $S_1$
Output: — $\lambda$ (1) (1) 00 (2) (1) (Contd.)

Input : $T(3)$ $I_{e2}$ $I_{e2}$ $T(4)$ $I_{e2}$ $I_{e2}$ $T(1)$
State : $S_1$ $S_3$ $S_1$ $S_1$ $S_4$ $S_1$ $S_1$ $S_1$
Output: 0 (3) (1) 00 (0) (1) $\lambda$ (Contd.)

Input : $I_{e1}$ $I_{e1}$ $T(2)$ $I_{e1}$ $I_{e1}$ $T(3)$ $I_{e1}$ $I_{e1}$
State : $S_1$ $S_1$ $S_1$ $S_2$ $S_1$ $S_1$ $S_3$ $S_1$ $S_1$
Output: (1) (1) 00 (2) (1) 0 (3) (1)
(Contd.)

Input : $T(4)$ $I_{e1}$ $I_{e2}$ $T(1)$ 0 $I_{e2}$ $T(1)$
State : $S_1$ $S_4$ $S_4$ $S_1$ $S_1$ $S_2$ $S_1$ $S_1$
Output: 00 (0) (0) $\lambda$ 1 (2) $\lambda$ (Contd.)

*Input : 1 $I_{e2}$ $T(1)$ $I_{e3}$ $I_{e2}$ $T(1)$ $I_{e4}$
State : $S_1$ $S_1$ $S_1$ $S_1$ $S_1$ $S_1$ $S_1$ $S_3$
Output: 1 (1) $\lambda$ 0 (1) $\lambda$ 0 (Contd.)

Input : $I_{e2}$ $T(2)$ 1 $I_{e2}$ $T(2)$ $I_{e3}$ $I_{e2}$
State : $S_3$ $S_1$ $S_2$ $S_3$ $S_1$ $S_2$ $S_1$ $S_1$
Output: (3) 00 0 (3) 00 1 (1) (Contd.)

Input : $T(2)$ $I_{e4}$ $I_{e2}$ $T(3)$ 0 $I_{e2}$ $T(3)$
State : $S_1$ $S_2$ $S_3$ $S_1$ $S_3$ $S_2$ $S_1$ $S_3$
Output: 00 1 (3) 0 0 (2) 0 (Contd.)

Input : 1 $I_{e2}$ $T(3)$ $I_{e3}$ $I_{e2}$ $(T3)$ $I_{e4}$
State : $S_3$ — $S_1$ $S_3$ $S_2$ $S_1$ $S_3$ $S_4$
Output: 1 (—) 0 0 (2) 0 0 (Contd.)

Input : $I_{e2}$ $T(4)$ $I_{e3}$ $I_{e2}$ $T(4)$ $I_{e4}$ $I_{e2}$
State : $S_4$ $S_1$ $S_4$ $S_2$ $S_1$ $S_4$ $S_4$ $S_1$
Output: (0) 00 1 (2) 00 1 (0)

The above input-output sequence confirms that the transition table of the modified machine is isomorphic to one described by Table 5.12. Thus it represents a complete checking sequence for the modified machine $M_2'$. The checking sequence also confirms that $I_{e1}$ is a distinguishing-cum-synchronizing sequence of length 1 for the original machine $M_2$, whereas $I_{e2}$ represents a synchronizing-cum-distinguishing sequence for the modified machine $M_2'$. All the transfer sequences of the machine $M_2'$ are verified in the process of checking the machine with the help of the distinguishing-cum-synchronizing sequences $I_{e1}$, and $I_{e2}$, respectively. In the above-mentioned checking sequence, the transfer sequence $T(1)$ for the state $S_1$ is shown specifically at different places, yet it never results in an increase in the length of the checking experiment, as the sequence $T(1)$ simply signifies a null sequence. The extra input symbol $I_{e2}$ alone can serve the function of distinguishing-cum-synchronizing sequence of length 1 and the utilization of another extra input symbol $I_{e1}$ seems to be unnecessary so far as the construction of the modified machine is concerned. So it becomes evident that the advantages of such types of modified machine can be achieved through the use of three extra input symbols $I_{e2}$, $I_{e3}$, and $I_{e4}$ only.

*In the above checking sequence, the presence of a dash (—) in the state and output sequence represents a don't care condition. Also, as before, any output which appears inside the parentheses represents the decimal value of the output appearing at the augmented output terminals.

The superiority of the present approach over that of Fujiwara et al. [85] results from the fact that the single application of $I_{e2}$ is possible here as compared to the $\lceil \log_2 q \rceil$ application $I_{e3}$ as a distinguishing-cum-synchronizing sequence. This reduces the length of the checking experiment and thus makes the machine 'very easily testable'.

## †5.6 Easily testable machines with extra inputs and extra outputs

In this section, we like to present yet another scheme of making an easily testable sequential machine by adding extra inputs and extra outputs as suggested by Das et al. [62]. Here also, the augmentation is affected by adding two extra input symbols and $\lceil \log_2 q \rceil$ extra output terminals, in the original machine $M$ having $q$ states and $p$ input symbols. For the sake of generality, the number of states $q$ satisfies the inequality $2^{m-1} < q < 2^m = q'$, where $m = \lceil \log_2 q \rceil$ is the total number of memory elements or state variables used in the realization of $M$. The following augmentation techniques are utilized:

(i) the new states $S_{q+1}, S_{q+2}, \ldots, S_{q'}$, are added to the machine $M$, where $q = 2^m$, and $m = \lceil \log_2 q \rceil$, the number of state variables;

(ii) an $m$ bit binary code is assigned to each state of $M$ so that each state gets one assignment only;

(iii) the two new input symbols $I_{e0}$, $I_{e1}$ and $m = \lceil \log_2 q \rceil$ output terminals $Z_1, Z_2, \ldots, Z_m$ are added to the machine $M$.

The next state functions $f$ corresponding to the new input symbols $I_{e0}$ and $I_{e1}$ are defined as follows: for each state $S_i$ of the augmented machine $M^*$ with a state assignment $y_1 y_2 \ldots y_m$, where a $y_i$, $i = 1, 2, \ldots, m$, represents a present state variable, $f(S_i, I_{e0}) = S_j$ and $f(S_i, I_{e1}) = S_k$, where the next states $S_j$ and $S_k$ have assignments $0Y_1Y_2 \ldots Y_{m-1}$ and $1Y_1Y_2 \ldots Y_{m-1}$, respectively, and a $Y_i$, $i = 1, 2, \ldots, m$ denotes a next state variable.

The output functions $g$ corresponding to the input symbols $I_{e0}$ and $I_{e1}$ are $g(S_i, I_{e0}) = i$, $1 \leqslant i \leqslant q - 1$; $g(S_r, I_{e0}) = 0$; $q + 1 \leqslant r \leqslant q'$; and $g(S_j, I_{e1}) = j$; $1 \leqslant j \leqslant q' - 1$; $g(S\ \ , I_{e1}) = 0$. The outputs corresponding to the added inputs would appear at the added output terminals, the decimal value of output being $\alpha$ and $\beta$, elements of the sets $\{0, 1, \ldots, q\}$ and $\{0, 1, 2, \ldots, q' - 1\}$, respectively.

The effect of this state transition is to shift the state assignment one digit to the right and to introduce a 0 or 1 as the leftmost digit according to the application of special inputs $I_{e0}$ and $I_{e1}$, respectively.

EXAMPLE: Consider the example of the machine $M_2$ as shown in Table 5.11, its modified machine $M_2^*$ according to the present scheme is shown in Table 5.13.

†Illustrations (Tables 5.13 etc.) in Section 5.6 are from the paper referred [62] in the bibliography and are presented with the permission of IEE (copyright © 1980 IEE).

Table 5.13 Machine $M_2^*$

| Present states | Next states and present outputs | | | |
|---|---|---|---|---|
| | $I_1 = 0$ | $I_2 = 1$ | $I_{e0}$ | $I_{e1}$ |
| $(00)S_1$ | $S_2$, 1 | $S_1$, 1 | $S_3$, 1 | $S_1$, 1 |
| $(01)S_2$ | — | $S_3$, 0 | $S_3$, 2 | $S_1$, 2 |
| $(10)S_3$ | $S_2$, 0 | —, 1 | $S_4$, 3 | $S_2$, 3 |
| $(11)S_4$ | — | — | $S_4$, 0 | $S_2$, 0 |

This machine $M^*$ has a distinguishing sequence $I_{e0}$ of length 1, with admissible set $A_1 = \{S_1, S_2, \ldots, S_q\}$ and a distinguishing sequence $I_{e1}$ of length 1 with admissible set $A_2 = \{S_1, S_2, \ldots, S_{q'}\}$, a synchronizing sequence of length $\lceil \log_2 q \rceil$ consisting of inputs $I_{e0}$ and $I_{e1}$, and similarly, transfer sequences of length that is at most $\lceil \log_2 q \rceil$ to move the machine from any arbitrary state to any of its other states. Thus the augmented machine $M^*$ with $q$ states and $(p + 2)$ input symbols becomes an easily testable sequential machine.

Checking sequence of this easily testable sequential machine can be designed following the principle of Fujiwara et al. as discussed earlier in Section 4.6 of Chapter 4. The initializing part as well as the validating part of the checking experiment remains the same as that of Fujiwara et al. [85], whereas the transition checking part needs an additional application of the synchronizing sequence to reset the machine to the initial state after checking each transition, because in the present case the distinguishing sequence is simultaneously not a synchronizing sequence. The total length $L$ of the checking experiment is thus at most

$$L = |\bar{I}_s| + \sum_{i=1}^{q} (|\bar{I}_{ti}| + |\bar{I}_d| + |\bar{I}_s|) + \sum_{j=1}^{p} \sum_{i=1}^{q} (|\bar{I}_{ti}| + |I_j| + |\bar{I}_d| + |\bar{I}_s|)$$

where $|\bar{I}_{ti}|$, $|\bar{I}_d|$, $|\bar{I}_s|$ represent the length of the initializing sequence, distinguishing sequence and synchronizing sequence respectively, or in other words,

$$L = (2q + 1)\lceil \log_2 q \rceil + q + pq(2\lceil \log_2 q \rceil + 2)$$

symbols, which is always less than the bound of Fujiwara et al. so long as $q \cdot (\lceil \log_2 q \rceil - 1 - p)$ is positive.

Additionally, in this approach also the faults that cause an increase in the number of states of the machine can also be detected. Suppose, on the application of a particular input symbol $I_j$, the transition leads the machine to a faulty state $S_x$, $q + 1 \leqslant x \leqslant q'$. Though the input $I_{e0}$ would produce an output 0 for state $S_x$, yet it can be identified by applying the input $I_{e1}$ only once and noting simultaneously the decimal value of the output being produced.

## †5.7 A novel approach for the checking experiments in sequential machines

In the last section of this chapter, we would discuss a method for designing efficiently the checking experiment for a synchronous sequential machine $M$ and for its fault diagnosis, by augmenting the said sequential machine $M$ with two extra input symbols $I_{e1}$ and $I_{e2}$, and with one extra output terminal $Z^*$ only. The method is an extension of the work of Fujiwara et al. [85] and has recently been developed by the author [25]. Here the outputs corresponding to the original inputs are observed at the original output terminals, while the outputs for the extra inputs $I_{e1}$ and $I_{e2}$ are observed at $Z^*$ terminal only. Like the other methods discussed earlier in this chapter, the original machine $M$ remains embedded inside the augmented machine $M^*$ and preserves its own input-output relationships with respect to the original input and original output terminals of the machine. Further, with respect to the application of special input symbols $I_{e1}$ and $I_{e2}$ and the output observed at the augmented output terminal $Z^*$, the behaviour of the machine would be like that of a Mealy model machine (or a pulse-input/pulse-output machine), although the original machine may be a Moore model machine (or a pulse-input/level-output machine). One major advantage of this method stems from the fact that there is no need to assume fault-free performance of the machine with respect to the next state and output transitions of the two special input symbols $I_{e1}$ and $I_{e2}$, since the developed procedure takes into account the correctness of the two special input symbols. Further, the experimental procedure clearly reveals the states of the machine in terms of the respective state assignments; thus at any instant the internal picture of the $m$ state variables can well be established from the terminal measurements performed on the machine.

### 5.7.1 *Scheme for machine modification*

It is assumed here that the machine is to be designed with exactly $m = \lceil \log_2 q \rceil$ secondary variables where $q$ denotes the number of states of the original machine $M$, and the creation of new internal variable is not at all possible. The original machine $M$, on which the augmentation would be affected, may or may not be a reduced, strongly connected and completely specified machine. Though the number of states $q$ of the original machine $M$ may be equal to an integral power of 2, yet for the sake of general discussion, it is assumed that the number of states $q$ of the machine satisfies the inequality $2^{m-1} < q < 2^m = q'$, where $m = \lceil \log_2 q \rceil$ is the total number of memory elements in the machine $M$, so that some of the $m$-tuples of the $m$-state variables remain unassigned to the states of

†Illustrations (equations, tables etc.) in Section 5.7 are from the paper referred [25] in the bibliography and are presented with the permission of IEE (copyright © 1983 IEE).

the original machine $M$. Now to achieve the modified machine $M^*$ of the original machine $M$ the following additional implementations are required to be incorporated to the original machine $M$.

(i) The new states $S_{q+1}, S_{q+2}, \ldots, S_{q'}$, are added to the machine $M$, where $q' = 2^m$.

(ii) An $m$-bit binary code is assigned to each state of $M^*$ so that only one $m$-tuple is assigned to each state and define the two states $S_1$ and $S_0$ as having all-one and all-zero combinations of $m$-state variables respectively.

(iii) The two new input symbols $I_{e1}$ and $I_{e2}$ are added to the machine $M$ and the next state and the output functions for these two special input symbols are defined as follows:

(a) For each state $S_i$ of $M$, with a state assignment $y_1y_2 \ldots y_m$, where $y_i$, $i = 1, 2, \ldots, m$ represents a present state variable, $f(S_i, I_{e1}) = S_j$ and $f(S_i, I_{e2}) = S_k$, where the next state $S_j$ and $S_k$ have assignments $0Y_1Y_2 \ldots Y_{m-1}$ and $1Y_1Y_2 \ldots Y_{m-1}$, respectively where $Y_i$, $i = 1, 2, \ldots, m$ denotes a next state variable. The effect of the state transitions corresponding to the added inputs is to shift the state assignment one digit towards the right, and to introduce thus a 0 or a 1 as the leftmost digit according as the input is $I_{e1}$ or $I_{e2}$, respectively.

(b) For any state $S_i$ having state assignments $y_1y_2 \ldots y_m$, the output mappings at the augmented output terminal $Z^*$ corresponding to the input symbols $I_{e1}$ and $I_{e2}$ are defined as $g(S_i, I_{e1}) = 0$ if $y_m = 0$ or 1, but $g(S_i, I_{e2}) = 0$ if $y_m = 0$ and $g(S_i, I_{e2}) = 1$ if $y_m = 1$, respectively.

The 2-column submachine restricted to the inputs $I_{e1}$ and $I_{e2}$ thus becomes isomorphic to an $m$-stage, 2-input symbol binary shift register with one extra output terminal $Z^*$. Now, we define the transfer sequences and the distinguishing sequence for the machine $M^*$ as follows:

(i) The transfer sequence or simply the $T_i$ sequence from the state $S_1$ to any state $S_i$ having state assignments $y_1y_2 \ldots y_m$, is defined to be an input sequence consisting of $m$ input symbols or either $I_{e1}$ or $I_{e2}$ applied as follows.

(a) Apply the first input symbol either $I_{e1}$ or $I_{e2}$ depending upon whether $S_i$ has the assignment for which $y_m = 0$ or $y_m = 1$, respectively, and (b) for any further application, say the $n$th one out of the remaining $(m - 1)$ applications, the choice of either $I_{e1}$ or $I_{e2}$ is made accordingly as the variable $y_{m-n+1}$ is 0 or 1 in the state assignment of $S_i$.

Thus for the modified machine $M^*$ we have the different $T_i$ sequences of length $m = \lceil \log_2 q \rceil$ symbols for every state $S_i$ of the machine which transfers the machine from state $S_1$ to the state $S_i$ for all $i$.

(ii) Similarly, the modified machine $M^*$ does possess a distinguishing sequence of length $m = \lceil \log_2 q \rceil$ symbols which is utilized to identify a state having assignment $y_1^*y_2^* \ldots y_m^*$ and also serves additionally to force the machine to the state $S_1$. This sequence having the distinguishing and synchronizing characteristics simultaneously has been called the $DI_{e2}$ sequence and is defined as follows:

It is an input-output sequence where the input sequence consists of $m$ repeated applications of $I_{e2}$ and the output sequence is defined as follows: (a) If the output is 1 at the very first application of $I_{e2}$, then $y_m^* = 1$; otherwise, the output 0 indicates $y_m^* = 0$; (b) for the rest $(m - 1)$ applications of $I_{e2}$ say the $n$th application would specify the variable $y_{m-n+1}^*$, variable 1 or 0 depending upon whether the output observed at the $n$th application is 1 or 0 respectively.

It is also evident from the above that after the $m$ repeated applications of $I_{e2}$ the machine $M^*$ synchronizes to the state $S_1$ having all-one combinations of $m$ state variables. Also, from the above discussion it is obvious that the modified machine $M^*$ stands for an easily testable sequential machine with built-in efficient fault detection capabilities as it contains (i) $DI_{e2}$ sequence which represents a distinguishing-cum-synchronizing sequence of length $m = \lceil \log_2 q \rceil$ symbols, and (ii) $T_i$ sequences of length $m = \lceil \log_2 q \rceil$ symbols for every state $S_i$ of the machine which transfers the machine from the state $S_1$ to the state $S_i$ for all $i$.

EXAMPLE: Consider again the example of a machine $M$ shown in Table 5.14; its modified version $M^*$ as per the present scheme of modification is shown in Table 5.15. The response of the various states of the above machine to the distinguishing sequence ($DI_{e2}$ sequence consisting of two repeated applications of $I_{e2}$ input symbols) is shown in Table 5.16(a); whereas the various transfer sequences with final states (for transferring the machine $M^*$ from the initial state $S_1$ to the various other states of the machine $M^*$) is shown in Table 5.16(b).

Table 5.14 The machine $M$

| Present states | Next states and present outputs | |
|---|---|---|
| | $I_1 = 0$ | $I_2 = 1$ |
| $(00)S_0$ | $S_2$, 1 | $S_0$, 1 |
| $(01)S_2$ | — | $S_3$, 0 |
| $(10)S_3$ | $S_2$, 0 | —, 1 |

Table 5.15 The machine $M^*$

| Present states | Next states and present outputs | | | |
|---|---|---|---|---|
| | $I_1 = 0$ | $I_2 = 1$ | $I_{e1}$ | $I_{e2}$ |
| $(00)S_0$ | $S_2$, 1 | $S_0$, 1 | $S_0$, 0 | $S_3$, 0 |
| $(01)S_2$ | — | $S_3$, 0 | $S_0$, 0 | $S_3$, 1 |
| $(10)S_3$ | $S_2$, 0 | —, 1 | $S_2$, 0 | $S_1$, 0 |
| $(11)S_1$ | — | — | $S_2$, 0 | $S_1$, 1 |

Table 5.16(a)

| Initial states | Response to $DI_{e2}$ sequence | Final states |
|---|---|---|
| $S_0$ | 00 | $S_1$ |
| $S_2$ | 10 | $S_1$ |
| $S_3$ | 01 | $S_1$ |
| $S_1$ | 11 | $S_1$ |

Table 5.16(b)

| Initial states | Transfer $(T_i)$ sequences | Final states |
|---|---|---|
| $S_1$ | $I_{e1}I_{e1}$ | $S_0$ |
| $S_1$ | $I_{e2}I_{e1}$ | $S_2$ |
| $S_1$ | $I_{e1}I_{e2}$ | $S_3$ |
| $S_1$ | $I_{e2}I_{e2}$ | $S_1$ |

### 5.7.2 *A novel approach of fault detection*

The following procedure has been adopted for developing a checking experiment for the machine $M$ contained in the modified machine $M^*$.

*Step 1*: The extra input symbol $I_{e1}$ is applied repeatedly $m$ times so that at the end of the operation the machine comes back to the state $S_0$.

*Step 2*: The extra input symbol $I_{e2}$ is applied repeatedly $2m$ times, and the output at the augmented output terminal $Z^*$ is observed. If the output is a concatenation of $m$ repeated symbols 0, followed by $m$ repeated symbols 1, then the performance of each of the extra input symbols $I_{e1}$ or $I_{e2}$ is correct and the machine would come back definitely to the state $S_1$ after such operation. If the desired output is not obtained the experiment cannot be conducted further and would indicate the incorrect operations of the special input symbols $I_{e1}$ and $I_{e2}$.

*Step 3*: For every state $S_i \in \{S_1, S_2, \ldots, S_{q'}\}$ of the modified machine $M^*$, the corresponding $T_i$ sequence is applied to set the machine to the state $S_i$ and at the end of such operation the $DI_{e2}$ sequence is applied to note the corresponding output at the extra output terminal $Z^*$. This finally would bring the machine to the state $S_1$ and would uniquely verify the transfer sequences to each state of the machine $M^*$ from the initial state $S_1$.

*Step 4*: To check the transition from any state $S_i$ belonging to the state alphabet of the original machine $M$ by an input symbol $I_j$ belonging to the input alphabet of the original machine $M$, the following subsequences are applied.

(i) The corresponding $T_l$ sequence is applied to set the machine to the state $S_i$.

(ii) An input $I_j$ is applied to the original input terminals and the output $O_j$ is observed at the original output terminals. This input $I_j$ would make the transition of the state $S_i$ to the state say, $S_r$.

(iii) The actual assignment of the state $S_r$ is then identified by applying the $DI_{e2}$ sequence and noting down the response at the augmented output terminal $Z^*$. This would finally bring back the machine to the state $S_1$.

Based on the knowledge of the aforesaid steps, the checking sequence of the machine $M$ would thus be designed quite easily. The entire checking sequence of $M$ can be split into a number of subsequences of which (a) the step 1 and step 2 would jointly require $3m$ input symbols, (b) the step 3 would need $q' \cdot 2m$ input symbols, and (c) if $p$ denotes the number of input symbols in the input alphabet of the machine $M$, then step 4 would require $p \cdot q \cdot (m + 1 + m)$ input symbols. Therefore, the length $L$ of the checking sequence would become

$$L = 3m + 2m \cdot q' + (2m + 1) \cdot p \cdot q \text{ symbols}$$

Now replacing, $q' = 2^m = 2^{\lceil \log_2 q \rceil}$ and $m = \lceil \log_2 q \rceil$ the above length $L$ can be written as,

$$L = \lceil 3 \log_2 q \rceil + 2 \lceil \log_2 q \rceil \cdot 2^{\lceil \log_2 q \rceil} + pq(2 \lceil \log_2 q \rceil + 1)$$

Referring to the section 4.6 of the chapter 4, this bound on the length of the checking experiment can be compared to the bound of Fujiwara et al. [85] which is say $L'$, where

$$L' = (3q + 1)\lceil \log_2 q \rceil + pq(2 \lceil \log_2 q \rceil + 1)$$

For the different values of $q$, the improvement in the length of the checking sequence as obtained in the method as compared to that of Fujiwara et al. [85] is found by obtaining the values of the following expression:

$$L' - L = (3q + 1)\lceil \log_2 q \rceil - \{3 \lceil \log_2 q \rceil + 2 \lceil \log_2 q \rceil \cdot 2^{\lceil \log_2 q \rceil}\}$$

Thus (a) for $q = 4$, the method offers a saving of 26–24 = 2 input symbols, (b) for say, $q = 8$ and $q = 16$, the present approach offers a saving of 75–57 = 18, and 196–140 = 56 input symbols respectively. In certain cases, where the number of states $q$ of the machine lies in between $2^{m-1}$ and $2^m$ (e.g., for $q = 6$), the upper bound of Fujiwara et al. and of this method may be the same, but in that case also the present approach offers an additional advantage because of its inherent fault diagnosis capability as follows.

The present technique possesses inherently an increased fault coverage performance and the faults associated with the next state transitions offer no special problem. Like anyother methods that we have discussed in this chapter, the faults resulting in an increase in the number of states of the original machine $M$ would also be detected. Suppose, for any state $S_i$ belonging to the state alphabet of the original machine $M$, the application of a particular input symbol to $S_i$ transfers the machine to a faulty state $S_x$, where $q + 1 \leqslant x \leqslant q'$. Then the faulty state $S_x$ would also be identified by the application of $DI_{e2}$ sequence and observing the response at the extra output terminal $Z^*$. Finally, this would bring back the machine to the initial state $S_1$ from the faulty state $S_x$. Thus this would enable us to evaluate the faulty transition table of the machine in the presence of faults associated in the next state transition of the original machine $M$.

Chapter 6

# Transitions Matrix Approach for the Measurement and Control of Synchronous Sequential Machines

## 6.1 Introduction

One of the basic problems in the study of sequential machines is to identify the state of the machine under the investigation. If the initial state of the machine is known, the response of the same can always be predicted; on the other hand, if it is not known, the response of the machine even to the specified applications of the inputs becomes unpredictable. Once the state of the sequential machine is identified, the behaviour of the machine under all future circumstances becomes predictable, and the usual steps may be taken to force the machine into various modes of operation at the will of the investigator. The former class of problems comes under the broad category usually termed as *measurement problems*, whereas the later problem is commonly known as *control problem* in sequential machines [92, 29].

Broadly speaking, the measurement problems include (a) the case of initial state identification or diagnosing problems, and (b) the case of terminal state identification known as homing problem of which the synchronizing problem is a special item. In these measurement problems, the input sequences to be applied to the machine are fixed in advance, so that the preset experiments are only conducted. Most of these measurement problems are solved by applying known input sequences to the machine and noting down the resulting output sequences. On the other hand, the control problem in sequential machine is concerned with finding the input sequences which take a given machine from a known initial state to any predesignated terminal state. In general, the control problem is a two-step adaptive process. Here the first step is a measurement problem that takes the machine from the unknown initial state to a predesignated intermediate state and is followed by a second step of applying a control sequence that takes the machine from the intermediate state to the desired final state.

A matrix approach to the solution of the above-mentioned measurement and control problems in synchronous sequential machines has become important of late. In recent papers, Das et al. [61, 63], instead of resorting to the conventional approach of using the transition table and the corresponding response tree, made use of the transition matrix representation of the machine and its higher order forms to solve the measurement and control problems in sequential machines. The procedure adopted by Das et al. is not only simple, but also very systematic and completely algorithmic, and thus lends itself to easy computer implementation. Before presenting the complete algorithmic approach of Das et al. [61, 63] let us discuss in the following section the implication of transition matrix and its higher order forms.

## *6.2 Transition matrix and higher order forms

A transition matrix is viewed as the mathematical counter part of the transition diagram in sequential machine. For a $q$-state machine $M$, the transition matrix composed of $q$ rows and $q$ columns and is denoted by $[\mathbf{M}]$. Let $\{S_1, S_2, \ldots, S_q\}$ be the state alphabet of $M$. Let $(i, j)$ entry of $[\mathbf{M}]$ (i.e., the entry common to the $i$th row and $j$th column of $[\mathbf{M}]$) be denoted by $b_{ij}$, if and only if, there exists an input that takes the machine $M$ from the state $S_i$ to the state $S_j$ in the state diagram of $M$, and is zero otherwise. For clarity, it is customary to attach the label of $k$th state $S_k$ to the $k$th row and column and refer to the row and column as 'row $S_k$' and 'column $S_k$' respectively.

EXAMPLE 6.1: The flow table and the transition matrix of a Mealy machine $M_1$ is shown below in Table 6.1 and equation (6.1) respectively.

Table 6.1 Machine $M_1$

| Present states | Next states and present outputs | |
|---|---|---|
| | $I = 0$ | $I = 1$ |
| $S_1$ | $S_2$, 0 | $S_4$, 0 |
| $S_2$ | $S_1$, 0 | $S_2$, 0 |
| $S_3$ | $S_4$, 1 | $S_1$, 0 |
| $S_4$ | $S_4$, 1 | $S_3$, 0 |

$$[\mathbf{M}_1] = \begin{matrix} & 1 & 2 & 3 & 4 \\ 1 & 0 & (0,0) & 0 & (1,0) \\ 2 & (0,0) & (1,0) & 0 & 0 \\ 3 & (1,0) & 0 & 0 & (0,1) \\ 4 & 0 & 0 & (1,0) & (0,1) \end{matrix} \tag{6.1}$$

*Illustrations (examples, tables and equations) in the section 6.2 and in the next section 6.3 are reprinted by permission of the publisher Elsevier Science Publishing Co., Inc. from the paper referred [61] in the bibliography. Copyright 1979 *Information Sciences* of Elsevier Science Publishing Co., Inc.

For a Mealy machine $M$, $b_{ij} = \Sigma(I_N, O_N)$, where the $I_N$ is the present input that takes $M$ from $S_i$ to $S_j$, and $O_N$ is its corresponding present output, whereas for a Moore machine $M$, $b_{ij} = \Sigma(I_N, O_{N+1})$, where the $I_N$ is the present input that takes $M$ from $S_i$ to $S_j$ as before, but $O_{N+1}$ its corresponding output is the next output, the summation being in either case over all such input-output pairs.

The transition matrix is useful in the determination of the paths and cycles in the transition diagram, in the classification of machine states, in finding the different submachines, in testing whether a machine in strongly connected or not, in finding the set of equivalent states of a machine and in machine identification [92]. The transition matrix is advantageous whenever the operations cannot be carried out successfully by a human investigator because of the limitations of observing the transition diagram visually for the complexity involved in the diagram. The concept of transition matrix can well be extended to higher orders by defining a multiplication operation of the matrices similar to that of ordinary matrices.

Let [**A**] be a $q \times q$ transition matrix with the $(i, j)$ entry $x_{ij}$, and [**B**] be an another $q \times q$ transition matrix with the $(i, j)$ entry $y_{ij}$, then [**C**] = [**A**][**B**] is also a $q \times q$ matrix of which the $(i, j)$ entry, $z_{ij}$, is

$$\begin{aligned} z_{ij} &= x_{i1}y_{1j} + x_{i2}y_{2j} + \ldots + x_{iq}y_{qj} \\ &= \sum_{i=1}^{q} x_{ik}y_{kj} \end{aligned}$$

Here the order of the factors in each product $x_{ik}y_{kj}$ must be preserved i.e., $x_{ik}y_{kj}$ is not necessarily equivalent to $y_{kj}x_{ik}$. It is to be noticed that the $(i, j)$ entry $z_{ij}$ of [**C**] vanishes if certain of the $x_{mn}$'s or $y_{rs}$'s are zero. Now if both [**A**] and [**B**] represent the transition matrix [**M**] of a $q$-state sequential machine, then the matrix [**C**] simply becomes equivalent to $[\mathbf{M}]^2$ with $(i, j)$ entry denoted by $b_{ij}^2$, where

$$b_{ij}^2 = \sum_{k=1}^{q} b_{ik}b_{kj}$$

Here also similarly, $b_{ij}^2$ vanishes if some of the $b_{mn}$'s are zero and $[\mathbf{M}]^2$ is referred to as the second order transition matrix. Therefore, for a $q$ state sequential machine $M$, a second order transition matrix is denoted by $[\mathbf{M}]^2$ and is composed of $q$ rows and $q$ columns, which are labeled in [**M**]. In general, the $(i, j)$ entry of the $r$th order transition matrix denoted by $[\mathbf{M}]^r$ is given as

$$b_{ij}^r = \sum_{k1,\, k2,\, \ldots,\, k(r-1)=1}^{q} b_{ik1}b_{k1k2} \ldots b_{k(r-1)j}$$

which, as usual, vanishes if some of the $b_{mm}$'s are zero. The rows and columns of $[\mathbf{M}]^r$ also labeled in [**M**]. Given a $k$th order transition matrix $[\mathbf{M}]^k$, the next higher order transition matrix $[\mathbf{M}]^{k+1}$ can be formed as $[\mathbf{M}]^{k+1} = [\mathbf{M}][\mathbf{M}]^k$.

EXAMPLE 6.2: The second order transition matrix $[\mathbf{M}_1]^2 = [\mathbf{M}_1][\mathbf{M}_1]$ corresponding to the machine $M_1$ of Table 6.1 is shown in equation 6.2.

$$[\mathbf{M}_1]^2 = \begin{array}{c} \\ 1 \\ 2 \\ 3 \\ 4 \end{array}\!\!\begin{array}{c} \begin{array}{cccc} 1 & 2 & 3 & 4 \end{array} \\ \begin{bmatrix} 0 & (0,0) & 0 & (1,0) \\ (0,0) & (1,0) & 0 & 0 \\ (1,0) & 0 & 0 & (0,1) \\ 0 & 0 & (1,0) & (0,1) \end{bmatrix} \end{array} \times \begin{array}{c} \\ 1 \\ 2 \\ 3 \\ 4 \end{array}\!\!\begin{array}{c} \begin{array}{cccc} 1 & 2 & 3 & 4 \end{array} \\ \begin{bmatrix} 0 & (0,0) & 0 & (1,0) \\ (0,0) & (1,0) & 0 & 0 \\ (1,0) & 0 & 0 & (0,1) \\ 0 & 0 & (1,0) & (0,1) \end{bmatrix} \end{array}$$

$$[\mathbf{M}_1]^2 = \begin{array}{c} \\ 1 \\ 2 \\ 3 \\ 4 \end{array}\!\!\begin{array}{c} \begin{array}{cccc} 1 & 2 & 3 & 4 \end{array} \\ \begin{bmatrix} (00,00) & (01,00) & (11,00) & (10,01) \\ (10,00) & (00,00)+(11,00) & 0 & (01,00) \\ 0 & (10,00) & (01,10) & (11,00)+(00,11) \\ (11,00) & 0 & (01,10) & (10,01)+(00,11) \end{bmatrix} \end{array} \tag{6.2}$$

It becomes obvious that the $(i, j)$ entry of $[\mathbf{M}_1]^2$ gives all input-output sequence $(\bar{I}_k, \bar{O}_m)$ of length two, such that $\bar{I}_k$ takes $M_1$ from the states $S_i$ to the state $S_j$ passing through some intermediate state $S_n$, not necessarily distinct from $S_i$ or $S_j$, producing the corresponding $\bar{O}_m$. In general, the $(i, j)$ entry of any $r$th order transition matrix $[\mathbf{M}]^r$ gives all input output sequences $(\bar{I}_k, \bar{O}_m)$ of length $r$, such that $\bar{I}_k$ takes $M$ from the state $S_i$ to the state $S_j$, passing through $(r - 1)$ intermediate states, not necessarily all distinct or distinct from $S_i$ or $S_j$, producing the corresponding $\bar{O}_m$. Also, it becomes at once clear from Eq. (6.2) that for any second order (or $r$th order transition matrix) no two entries in the same row can have terms that involve the same input sequence $\bar{I}_k$.

## 6.3 State identification by transition matrix approach

Given a transition table of a $q$-state synchronous sequential machine $M$, the corresponding transition matrix $[\mathbf{M}]$ of the same and its higher order forms may be formed and the various state identification in sequential machines may be carried out.

### 6.3.1 *Terminal state identification for homing problem*

THEOREM 6.1: An input sequence $\bar{I}_n$ of length $r$ is a homing sequence for a synchronous sequential machine $M$, if and only if, in the $r$th order transition matrix $[\mathbf{M}]^r$ of $M$, whenever $\bar{I}_n$ appears in the entries of two or more columns which are all distinct, the corresponding output sequences of $\bar{I}_n$ in the entries of those columns are also distinct.

*Proof*: Let us take that the entry in row $i$ and column $j$ of $[\mathbf{M}]^r$ is $(\bar{I}_n, \bar{O}_l)$, and the entry in a row $m$ and column $n$ is also $(\bar{I}_n, \bar{O}_l)$, but $j \neq n$. It signifies that the input sequence $\bar{I}_n$ can take the machine from the state $S_i$ to $S_j$ or from the state $S_m$ to $S_n$, producing the same output sequence $\bar{O}_l$. Therefore, by observing the output response $\bar{O}_l$ after the application of the input sequence $\bar{I}_n$, no one can distinguish between the states $S_j$ and $S_n$. Indirectly, thus it becomes obvious that in order to distinguish between $S_j$ and $S_n$ the output sequences produced in response to the application of $\bar{I}_n$ to $M$ in the respective states $S_i$ or $S_m$ have to be distinct, i.e. the corresponding output sequences of $\bar{I}_n$ in the entries of columns $j$ and $n$ have to be distinct. This must be true for every pair of distinct columns of $[\mathbf{M}]^r$ where $\bar{I}_n$ appears in the entries. This proves the theorem.

### 6.3.2 *Terminal state identification for synchronizing problem*

THEOREM 6.2: An input sequence $\bar{I}_n$ of length $r$ is a synchronizing sequence for a $q$ state sequential machine $M$, if an only if, in the $r$th-order transition matrix $[\mathbf{M}]^r$ of $M$, $\bar{I}_n$ appears in the entries of all the rows for a particular column $j$, $1 \leqslant j \leqslant q$.

*Proof*: An input sequence $\bar{I}_n$ of length $r$ becomes a synchronizing sequence whenever $f(\bar{I}_n, S)$ would become a mapping of the state set $S$ of $M$ on to a single state $S_j$ belonging to state set $S$. This necessitates that in the $r$th order transition matrix $[\mathbf{M}[^r$ of $M$, in column $j$, $1 \leqslant j \leqslant q$, $\bar{I}_n$ should appear in the entries of all the rows which proves evidently the theorem.

### 6.3.3 *Initial state identification for diagnosing problem*

THEOREM 6.3: An input sequence $\bar{I}_n$ of length $r$ becomes a distinguishing sequence for a sequential machine $M$ with admissible set $A(M) = S$, the set of states of $M$, if and only if, in the $r$th order transition matrix $[\mathbf{M}]^r$ of $M$, for all of the $q$ rows in each of which $\bar{I}_n$ appears in the entry of some column $j$, $1 \leqslant j \leqslant q$, the corresponding output sequences of $\bar{I}_n$ in the entries of all the columns are distinct.

*Proof*: Suppose the entry in row $i$ and column $j$ of $r$th order transition matrix $[\mathbf{M}]^r$ is $(\bar{I}_n, \bar{O}_l)$ and that in the row $m$ and column $n$ is also $(\bar{I}_n, \bar{O}_l)$ where $i \neq m$ and $j$ are and $n$ not necessarily distinct. This indicates that the input sequence $\bar{I}_n$ takes the machine $M$ from the state $S_i$ to the state $S_j$, and also moves the machine from the state $S_m$ to the state $S_n$ but produces the output sequence $\bar{O}_l$ in both the cases. Therefore, it is impossible to distinguish between the initial states $S_i$ and $S_m$ of the machine $M$ by noting the resultant output responses thus produced. Thus, for distinguishing between the states $S_i$ and $S_m$, the output responses produced in response to the application of $\bar{I}_n$ to $M$ in the respective states must have to be distinct, i.e., the corresponding output sequences of $\bar{I}_n$ in the entries of columns $j$ and $n$ must have to be distinct. It should hold valid for every pair of distinct rows of $[\mathbf{M}]^r$ for which $\bar{I}_n$ must appear in the entries of certain columns. Thus the theorem is verified.

EXAMPLE 6.3: To illustrate these theorems let us take the example of the sequential machine $M_2$ shown in Table 6.2.

Table 6.2 The machine $M_2$

| Present states | Next states and present outputs | |
|---|---|---|
| | $I = 0$ | $I = 1$ |
| $S_1$ | $S_2$, 0 | $S_1$, 0 |
| $S_2$ | $S_2$, 1 | $S_3$, 1 |
| $S_3$ | $S_1$, 1 | $S_4$, 0 |
| $S_4$ | $S_3$, 0 | $S_1$, 1 |

The transition matrix of the above machine $M_2$ is shown below in equation 6.3.

$$[\mathbf{M}_2] = \begin{array}{c} \\ 1 \\ 2 \\ 2 \\ 4 \end{array} \begin{array}{c} \begin{array}{cccc} 1 & 2 & 3 & 4 \end{array} \\ \begin{bmatrix} (1, 0) & (0, 0) & 0 & 0 \\ 0 & (0, 1) & (1, 1) & 0 \\ (0, 1) & 0 & 0 & (1, 0) \\ (1, 1) & 0 & (0, 0) & 0 \end{bmatrix} \end{array} \tag{6.3}$$

The second and third order matrices of the abovementioned transition matrix $[\mathbf{M}_2]$ are shown in the equation 6.4 and 6.5 respectively.

$$[\mathbf{M}_2]^2 = \begin{array}{c} \\ 1 \\ 2 \\ 3 \\ 4 \end{array} \begin{array}{c} \begin{array}{cccc} 1 & 2 & 3 & 4 \end{array} \\ \begin{bmatrix} (11, 00) & (10, 00)+(00, 01) & (01, 01) & 0 \\ (10, 11) & (00, 11) & (01, 11) & (11, 10) \\ (01, 10)+(11, 01) & (00, 10) & (10, 00) & 0 \\ (11, 10)+(00, 01) & (10, 10) & 0 & (01, 00) \end{bmatrix} \end{array} \tag{6.4}$$

From the second order transition matrix $[\mathbf{M}_2]^2$ of the machine $M_2$ as given in equation (6.4) it becomes clear that the input sequence 01 of length two is a homing sequence for the machine. Here whenever 01 appears in the entries of two or more columns which are distinct in $[\mathbf{M}_2]^2$, the corresponding output sequences of those columns are also distinct. Also, upon the application of the corresponding homing sequence $\bar{I}_n$(i.e., 01) if the corresponding output sequence $\bar{O}_l$ would become 01, 11, 10 and 00 respectively, then the terminal state to which the machine would reach be $S_3$, $S_3$, $S_1$, and $S_4$, respectively.

$$
[\mathbf{M}_2]^3 = \begin{array}{c} \\ 1 \\ \\ 2 \\ \\ 3 \\ \\ 4 \end{array}
\overset{\begin{array}{cccc} 1 & \qquad 2 & \qquad 3 & \qquad 4 \end{array}}{\left[\begin{array}{rrrr}
\begin{array}{r}(111,000)\\+(010,011)\end{array} & \begin{array}{r}(110,000)\\+(100,001)\\+(000,011)\end{array} & \begin{array}{r}(101,001)\\+(001,011)\end{array} & (011,010) \\
\begin{array}{r}(010,111)\\+(101,110)\\+(111,101)\end{array} & \begin{array}{r}(000,111)\\+(100,110)\end{array} & \begin{array}{r}(001,111)\\+(110,100)\end{array} & (011,110) \\
\begin{array}{r}(011,100)\\+(111,010)\\+(100,001)\end{array} & \begin{array}{r}(010,100)\\+(000,101)\\+(110,010)\end{array} & (001,101) & (101,000) \\
\begin{array}{r}(111,100)\\+(001,010)\\+(011,001)\end{array} & \begin{array}{r}(110,100)\\+(100,101)\\+(000,010)\end{array} & \begin{array}{r}(101,101)\\+(010,000)\end{array} & 0
\end{array}\right]}
\tag{6.5}
$$

It also becomes at once apparent from equation (6.3) and (6.4) that the machine can be brought back to a state say $S_2$ from the state $S_1$, by applying the control sequence of length one (i.e., 0 for $S_1$); also even if the final states are $S_3$ and $S_4$, the machine may still be brought back to the state $S_2$ by applying the control sequence 00 and 10 to the respective states.

From the third order transition matrix $[\mathbf{M}_2]^3$ shown in equation (6.5) it becomes also evident that the input sequence $\bar{I}_n = 111$ or $\bar{I}_n = 000$ of length three are the two synchronizing sequences of the machine $M_2$. Because $\bar{I}_n = 111$ appears in the entries of all the four rows of the column $S_1$ of $[\mathbf{M}_2]^3$ and that of $\bar{I}_n = 000$ also appears in the entries of all the rows of the columns $S_2$ of $[\mathbf{M}_2]^3$. As this condition does not hold good for any other input sequences of length three for the same machine, the sequences 111 and 000 would become the only two synchronizing sequences of length three for the machine $M_2$.

From the second-order transition matrix $[\mathbf{M}_2]^2$ of equation (6.4) it becomes further clear that the input sequence $\bar{I}_n = 01$ of length two is not only the homing sequence of the machine but also stands for the distinguishing sequence of the same machine. Here in each of the four rows of $[\mathbf{M}_2]^2$ where 01 appears, the corresponding output sequence of 01 is distinct from that of 01 in all others. Further, it becomes obvious from equation (6.4) that on the application of input sequence $\bar{I}_n = 01$ if the output responses would become $\bar{O}_l = 01, 11, 10,$ and $00$, then the initial states of machine $M_2$ are $S_1$, $S_2$, $S_3$, and $S_4$, respectively.

Similarly, inspecting the entries of the different rows and columns of the third order transition matrix $[\mathbf{M}_2]^3$ of equation (6.5) it becomes evident that each of the individual sequences of length three (viz., 000, 001, 010, 011, 101, and 111) may also be taken as the preset distinguishing sequence for the machine $M_2$.

## *6.4 Multiple preset experiments for initial state identification

When only a single copy of the given machine is available there remains no way in general, of knowingly recovering the initial state of the machine for conducting a new experiment, in case the previous experiment proves to be a failure. If however, a sufficient number of copies of the given machine are available then it becomes possible often to conduct a number of experiments, each of which by itself may not be able to solve the initial state identification or diagnosing problem, but all of which jointly may supply us with the information to identify the initial state of the machine. For example, consider the case of the machine $M_1$ shown in Table 6.1. There is no simple experiment which solves the diagnosing problem for the same machine $M_1$ with admissible set $\{S_1, S_2, S_3, S_4\}$. But this problem is readily solvable by multiple preset experiment with multiplicity 2 and length 3.

In solving the diagnosing problem of sequential machines which do not possess distinguishing sequences, we resort to multiple preset experimentation on the copies of the same machine and we require to find a set of input sequences $\bar{I}_{k\beta}$, $\beta = 1, 2, \ldots, h$, such that there exists a unique relationship between the observed output sequences $g(S_j, \bar{I}_{k\beta})$, $\beta = 1, 2, \ldots, h$ and the unknown initial state of the machine $S_j$. It was discussed in Chapters 2 and 3 that in every minimal machine $M$, even if, $M$ does not possess a single diagnosing input sequence, it does always possess a set of input sequences, termed as characterizing sequences that can well be used for the identification of the unknown initial state of the machine. Also, the graphical exhaustive search procedure, which evaluates the responses through the successor tree formation for finding out the different characterizing set of the machine sometimes become too much complicated. For the same a new approach has been developed recently by Das et al. [63] that makes use of the concept of the transition matrix and its higher order forms.

A set of input sequences $\bar{I}_{k1}, \bar{I}_{k2}, \ldots, \bar{I}_{kh}$ of lengths $r_1, r_2, \ldots, r_h$, respectively, represents a characterizing sequence for a sequential machine $M$ with admissible set $A(M) = S_{a1}, S_{a2}, \ldots, S_{ak} \subseteq S$, the set of states of $M$, provided in each of the $a\theta$ rows, $\theta = 1, 2, \ldots, k$, of all of the $r_\alpha$-th order transition matrices $[\mathbf{M}]^{r_\alpha}$, $\alpha = 1, 2, \ldots, h$, wherever the input sequences $\bar{I}_{k\beta}$, $\beta = 1, 2, \ldots, h$, appear in the entries of some column $j$, $1 \leqslant j \leqslant q$, the corresponding output sequences of $\bar{I}_{k\beta}$, $\beta = 1, 2, \ldots, h$, in the entries of all those columns of all the matrices constitute a distinct set. To put it indirectly in other words, a set of input sequences $\bar{I}_{k1}, \bar{I}_{k2}, \ldots, \bar{I}_{kh}$ of lengths $r_1, r_2, \ldots, r_h$, respectively, represents a characterizing sequence for a sequential machine $M$ with admissible set $A(M) = S_{a1}, S_{a2}, \ldots, S_{ak} \subseteq S$, the set of states of $M$, if and only if, $\bar{I}_{k1}, \bar{I}_{k2}, \ldots, \bar{I}_{kh}$ define partitions $\pi_1, \pi_2, \ldots, \pi_h$ based on the output responses for the set of admissible states $A(M)$ in the matrices $[\mathbf{M}]^{r_1}, [\mathbf{M}]^{r_2}, \ldots, [\mathbf{M}]^{r_h}$, respectively, such that the

*Illustrations in the section 6.4 are from the paper referred [63] in the bibliography and are presented with the permission of IEE (copyright © 1979 IEE).

product of these partitions $\pi_1, \pi_2, \ldots, \pi_h = \phi$, would become the trivial partition in which there would be *ak* blocks, with each block containing exactly one state of $A(M)$.

This becomes at once apparent from the definition of characterizing sequence given in Chapter 2 and also from Theorem 6.3 stated earlier. As an example, consider the transition matrix $[\mathbf{M}_1]$ given in equation (6.1). If we select the input symbol 0, at once it becomes clear from $[\mathbf{M}_1]$ that by observing the response the admissible uncertainties regarding the initial states breaks up to the partitions $\overline{S_1, S_2}; \overline{S_3, S_4}$. Thus we can further select an input sequence 10 of length two, and it becomes also clear from the second order transition matrix $[\mathbf{M}_1]^2$ of equation (6.2) that it breaks the uncertainties to the trivial partition $\bar{S}_1, \bar{S}_2, \bar{S}_3, \bar{S}_4$. Therefore, {0, 10} stands in together for the characterizing sequence of the machine $M_1$ with admissible set $\{S_1, S_2, S_3, S_4\}$ of the machine.

## 6.5 Procedure of obtaining various state identification sequences

The entire approach depicted earlier may be generalized in an algorithm for the measurement problems in synchronous, Mealy model, deterministic sequential machines [61, 63]. The procedure is given as follows:

*Step 1*: Given the transition table of a minimal $q$-state sequential machine, its transition matrix [**M**] is formed.

### A. Homing Problem

*Step 2A*: The possibility for the existence of any homing sequence is ascertained according to the requirements of Theorem 6.1 by checking the entries of the different rows and columns in [**M**].

If it finds the same, then the homing sequence (sequences) are obtained and the various terminal states are also identified which stops the process; otherwise, it leads from here to next *Step 3A*.

*Step 3A*: Successively, the higher order matrices $[\mathbf{M}]^2 = [\mathbf{M}]\cdot[\mathbf{M}]$, $[\mathbf{M}]^3 = [\mathbf{M}]\cdot[\mathbf{M}]^2, \ldots,$ are formed and in each case, the *Step 2A* is repeated, unless for some $[\mathbf{M}]^r$, the process terminates resulting in the desired homing sequence (sequences). The process must end for some $[\mathbf{M}]^r$ with $r \leqslant (q-1)^2$. [Refer to Section 2.72 of Chapter 2.]

### B. Synchronizing Problem

*Step 2B*: The possibility of the existence of any synchronizing sequence is ascertained according to the requirements of Theorem 6.2 by checking the entries of the different rows and columns of [**M**].

If it does possess the same, then synchronizing sequence (sequences) are found out and the terminal state (or states) of the machine $M$ at which the machine would synchronize have also been identified, and the process stops; otherwise, it leads to next *Step 3B*.

*Step 3B*: Successively, the higher order matrices $[\mathbf{M}]^2 = [\mathbf{M}]\cdot[\mathbf{M}]$; $[\mathbf{M}]^3 = [\mathbf{M}]\cdot[\mathbf{M}]^2, \ldots$ are formed and in each case the *Step 2B* is repeated unless for some $[\mathbf{M}]^r$ the process terminates. The process would be terminating after a finite number of iterations for some $[\mathbf{M}]^r$, $r \leqslant q(q-1)^2/2$ if the machine **M** possesses any synchronizing sequence. [Refer to Section 2.73 of Chapter 2.] If however on going up to $[\mathbf{M}]^r$, for the value of $r$ as given above, the process does not terminate, then the machine **M** is said to have no synchronizing sequence and the entire process thus stops.

C. Diagnosing Problem I

*Step 2C*: The possibility of the existence of any distinguishing sequence is ascertained according to the requirements of Theorem 6.3 by checking the entries of the different rows and columns of $[\mathbf{M}]$.

If it does exist then the distinguishing sequence (or sequences) are found out and the initial states of machine are found out and the process stops; otherwise, (ii) if it does not possess any distinguishing sequence then it leads to the next *Step 3C*.

*Step 3C*: Successively, the higher order matrices $[\mathbf{M}]^2 = [\mathbf{M}]\cdot[\mathbf{M}]$, $[\mathbf{M}]^3 = [\mathbf{M}]\cdot[\mathbf{M}]^2, \ldots,$ are formed and in each case the *Step 3A* is repeated. In case the machine $M$ possesses distinguishing sequence then the process terminates after a finite number of iterations for some $[\mathbf{M}]^r$, $r \leqslant (q-1)q^q$ [Refer Section 2.71 of Chapter 2]. If however on leading up to $[\mathbf{M}]^r$, for $r$ as given above, the process does not terminate, then the machine $M$ does not possess any distinguishing sequence and the entire process of initial state identification stops.

D. Diagnosing Problem II

In case the machine $M$ does not possess distinguishing sequence (or sequences) then the characterizing sequences may be found out.

*Step 2D*: By checking the entries in the different rows and columns of the transition matrix $[\mathbf{M}]$ an input sequence $\bar{I}_s$ is selected such that with $\bar{I}_s$ a partition $\pi$ may be defined on the admissible set of states $A(M)$ of the machine $M$, having maximum number of blocks where each block $b_i$ corresponds to some distinct output. For any machine $M$ with the size of its output alphabet being $l$, the number of such partition blocks $n_i$ is always less than or equal to $l$.

*Step 3D*: The second order transition matrix $[\mathbf{M}]^2 = [\mathbf{M}]\cdot[\mathbf{M}]$ is formed.

*Step 4D*: By checking the entries in the different rows and columns of $[\mathbf{M}]^2$, again an input sequence is selected such that it partitions the state sets of a maximum number of blocks in $\pi$ and simultaneously for each such block $b_i$, it partitions its set of elements into a maximum number of disjoint subsets, based on the output responses. If after doing this, every block $b_j$ of $\pi$ becomes partitioned into single element sets, then the process stops.

*Step 4E*: Otherwise, successively, the higher order transition matrices $[\mathbf{M}]^3 = [\mathbf{M}]\cdot[\mathbf{M}]^2, \ldots,$ are formed and in each case the *Step 4D* is repeated, until for some $[\mathbf{M}]^r$ the process terminates and thus the desired characterizing sequence (or sequences) of the machine are obtained.

# Chapter 7

# Application of Checking Experiments

## 7.1 Introduction

In this chapter, we consider the fault detection in random access memories and in bit-sliced microprocessors. A memory fault model for the random access memory has been provided by J. P. Hayes [104], where functionally the behaviour of the memory has been made equivalent to that of an incompletely specified synchronous sequential machine and the checking experiments are directly conducted for the detection of failures in the memory. Similarly, a functional fault model for the bit-sliced microprocessor has recently been developed by T. Sridhar and J. P. Hayes [211] where the testing of the sequential modules inside the processor slices are carried out by developing suitable checking experiments for them. We shall now present in this chapter the wonderful approaches of fault detection in random access memories and in bit-sliced microprocessors.

## *7.2 Random access memory

Today, memory systems of large storage capacity comprising of semiconductor random access memories (RAM) are available and used in the computer main frames. Failures may occur in such memory systems due to various reasons. The physical causes of failure in semiconductor memory may depend on various factors like component density, circuit layout, and method of manufacture. By applying standard test patterns to the memory, certain faults may be detected. These methods test the READ and WRITE operation of each cell of the memory under the assumption that the results are generally affected by the data stored in certain other cells of the memory. Pattern sensitive faults (PSF) for a cell in the random access memory refer to the faults due to the neighbouring cells for their close proximity with the cell under consideration. There are many physical failure modes that produce pattern sensitivity. For example, electrical signals on conductors in close spatial proximity can interact with one another. In large scale integrated circuits, the high component and

*Illustrations (tables and figures etc.) in Section 7.2 are from the paper referred [104] in the bibliography and are presented with the permission of IEEE (copyright © 1975 IEEE).

connection densities aggravate this problem. Other instance of the pattern sensitivity results from the failure of a device to recognize a single 0 (or 1) that follows a long sequence of 1's (or 0's) in a particular line. This is exactly the case of time dependent PSF and it results due to unwanted hysteresis effects. A variety of heuristic procedures have been developed to detect pattern sensitive faults in memories. The most common test procedures are 'Write and Read Ones and Zeros', 'Marching Ones and Zeros', 'Walking Ones and Zeros' and 'Galloping Ones and Zeros' [14, 104]. In the memory of $r$ cells 'Walking 1/0 (ripple) test' requires approximately $2r^2 + 3r$ READ and WRITE operations, whereas 'Galloping 1/0 test' or widely used 'Galpat test' requires about $4n^2$ patterns to check $n$-bit of RAM. Further since most of PSF tests were derived empirically, their underlying fault models are rather difficult for understanding and this makes it further difficult to determine their fault coverage.

Various methods of testing pattern sensitive faults (PSF) for the RAM memories just enumerated have important features that they do not make use of the information about the circuitry from which the memory system has been constructed, i.e., the memory is considered to be 'black box' with accessible input/output terminals only. Thus the approach of testing resembles 'the machine identification' philosophy rather than the circuit testing approach. Also, in the above approaches of fault finding, no attempts have been made to formulate the logical models for the pattern sensitive faults, corresponding to the essentially logical nature of tests. Without such models, there was difficulty for evaluating analytically the fault coverage of a test procedure and whether the number of tests becomes optimal or near optimal. It was Hayes [104], who first treated this problem to formulate the general logical model for pattern sensitive faults in random access memories and presented a complete test procedure for finding the faults of various types. We shall now present the memory fault model given by J. P. Hayes and show how it converts a memory $M_r$ to an incompletely specified $q$-state Mealy model machine and subsequently discuss the procedure of fault detection in random access memories.

### 7.2.1 *Memory fault model*

Here we define a memory $M_r$ to be a set $\{C_0, C_1, \ldots, C_{r-1}\}$ of $r$ binary storage cells, e.g., flipflops. The subscript $i$ of $C_i$ is its address. READ/WRITE, these are the only two types of operation that can be performed upon any cell of this kind. For the sake of convenience, it is assumed that READ or WRITE operation can only refer to a single cell of the memory at any discrete time instant, i.e., the memory $M_r$ has been considered for a word size of 1 bit only. Since $M_r$ stands for a random access memory, any cell may be selected for reading or writing independently of the previous READ or WRITE operations.

Let two binary elements 0, 1 form a set $B = (0, 1)$ and $B^r$ be the set of all $2^r$ vectors with $r$ binary components. $M_r$ has $q = 2^r$ distinct states and each state $S_j$ is represented by a vector $(y_{j,0}, y_{j,1}, \ldots, y_{j,r-1})$ in $B_r$.

Now the three functions $W_i$, $\overline{W_i}$ and $R_i$, respectively, for writing 1, 0 and reading operation can be associated on the states of $M_r$ as follows:

$$f(S_j, W_i) = f\{(y_{j,0}, y_{j,1}, \ldots, y_{j,i}, \ldots, y_{j,r-1}), W_i\}$$
$$= (y_{j,0}, y_{j,1}, \ldots, 1, \ldots, y_{j,r-1}) \quad (7.1)$$
$$f(S_j, \overline{W_i}) = f\{(y_{j,0}, y_{j,1}, \ldots, y_{j,i}, \ldots, y_{j,r-1}), \overline{W_i}\}$$
$$= (y_{j,0}, y_{j,1}, \ldots, 0, \ldots, y_{j,r-1}) \quad (7.2)$$
$$f(S_j, R_i) = S_j \quad (7.3)$$

Suppose, $W_i$, or $\overline{W_i}$ is denoted by a single symbol $\hat{W}_i$ and $X_i$ is used to denote $W_i$, $\overline{W_i}$ or $R_i$, then we define an output function $g$ as follows:

$$g(S_j, X_i) = \begin{cases} y_{j,i} & \text{if } X_i = R_i \\ \text{`—'} & \text{if } X_i = \hat{W}_i \end{cases} \quad (7.4)$$

where '—' symbol stands for an unspecified or don't care value.

From the equations (7.1), (7.2), (7.3) and (7.4) it is obvious that $M_r$ can be considered as an incompletely specified $q$-state Mealy Model sequential machine with $3r$ input symbols and 2 output symbols. The inputs of $M_r$ are $W_i$, $\overline{W_i}$, $R_i$ for $i = 0, 1, \ldots, r-1$. The next state functions are given by the equations (7.1), (7.2) and (7.3), and the output function is denoted by the equation (7.4).

EXAMPLE: Let us consider a memory $M_2$ consisting of two cells only. Table 7.1 shows the corresponding flow table with $q = 2^2 = 4$ states and Fig. 7.1 shows the transition diagram $G_2$ of $M_2$.

Table 7.1 Machine $M_2$

| Present states | Next states and present outputs | | | | | |
|---|---|---|---|---|---|---|
| | $W_0$ | $W_1$ | $\overline{W_0}$ | $\overline{W_1}$ | $R_0$ | $R_1$ |
| $y_0$ $y_1$ | $Y_0$ $Y_1$ | $Y_0$ $Y_1$ | $Y_0$ $Y_1$ | $Y_0$ $Y_1$ | $Y_0$ $Y_1$ | $Y_0$ $Y_1$ |
| 0 0 ($S_0$) | 1 0,— | 0 1,— | 0 0,— | 0 0,— | 0 0, 0 | 0 0, 0 |
| 0 1 ($S_1$) | 1 1,— | 0 1,— | 0 1,— | 0 0,— | 0 1, 0 | 0 1, 1 |
| 1 0 ($S_2$) | 1 0,— | 1 1,— | 0 0,— | 1 0,— | 1 1, 1 | 1 0, 0 |
| 1 1 ($S_3$) | 1 1,— | 1 1,— | 0 1,— | 1 0,— | 1 1, 1 | 1 1, 1 |

It is obvious that the behaviour of $M_r$ is completely determined by the functions $F_0 = (W_i, \overline{W_i}, R_i, g)$, where $i = 0, 1, \ldots, r-1$. An unrestricted pattern sensitive fault is said to occur when the functions $(W_i, \overline{W_i}, R_i, g)$ change to $(W_i^F, \overline{W_i}^F, R_i^F, g^F)$, where $W_i^F$, $\overline{W_i}^F$, $R_i^F$ are arbitrary mappings on $B^r$, and $g^F$ is any mapping from $B^r$ to $B$, and the machine $M_r$ changes to its faulty model $M_r^F$ due to fault $F = (W_i^F, \overline{W_i}^F, R_i^F, g^F)$.

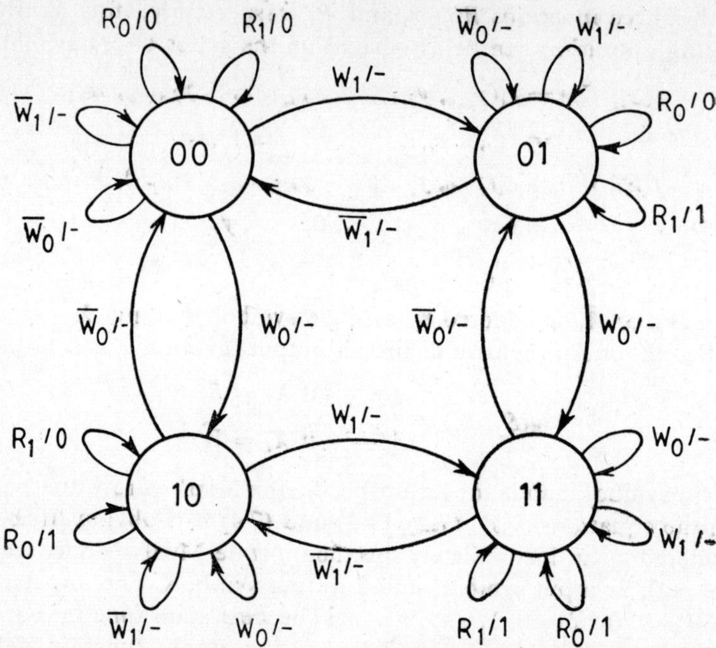


Fig. 7.1 Transition diagram $G_2$ for a memory $M_2$ consisting of two cells only.

### 7.2.2 *Mathematical representations for detectable and undetectable PSF*

Let us investigate the three important aspects of the 'state' in $M_r^F$. The internal state of $M_r^F$ at any time is the actual pattern of 0's and 1's stored in $M_r^F$ and is denoted by a vector of the form $S_j = (y_{j,0}, y_{j,1}, \ldots, y_{j,r-1})$. Let $\lambda$ be an input sequence, which when applied to $M_r$, leaves it in a unique final state $S_k$ independent of the initial state. Every sequence containing at least one WRITE operation on every cell has this property. Suppose $\lambda$ is applied to $M_r^F$ and let $S_j$ be the resulting internal state of $M_r^F$ at some time instant $t$. The state $S_j$ may depend on the initial state of $M_r^F$. Now if $M_r^F$ would have been fault free then it would have entered to a state say $S_k$ at that time, called the expected state of $M_r^F$. It is convenient to consider the expected state of $M_r^F$ to be a function of its internal state and denote it by $E(S_j)$. Therefore, for this $M_r^F$ at time instant $t$, we can write $E(S_j) = S_k$. Finally, the apparent state of $M_r^F$ at time instant $t$ is defined as $[g^F(S_j, R_0), g^F(S_j, R_1), \ldots, g^F(S_j, R_{r-1})]$ and is denoted by $A(S_j)$. This shows that if we attempt to READ cell $C_i$ of $M_r^F$ at time instant $t$, the output observed would be the value of $i$th component of $A(S_j)$.

From the above discussion it becomes clear that if the faulty $M_r^F = M_r$, then all the three types of states coincide. The fault $F$ is called a detectable PSF if and only if for some internal state $S_j$ of $M_r^F$ at some time instant $t$, $E(S_j) \neq A(S_j)$. On the other hand, the fault $F$ is said to be undetectable PSF if $E(S_j) = A(S_j)$ at all times, but $S_j$ might differ from $E(S_j)$ and $A(S_j)$.

### 7.2.3 *Features of the checking experiment*

We have seen earlier the memory model $M_r$, which is definitely a strongly connected sequential machine. Since unrestricted PSF permit $M_r$ to change to its faulty model $M_r^F$ with $q = 2^r$ or fewer states, the problem resolves to that of deriving a checking sequence for $M_r$. In view of the unspecified outputs of $M_r$, the checking sequence for $M_r$ need only to distinguish $M_r$, from all the machines which are incompatible to $M_r$. Thus a change in the unspecified value of $M_r$, while conducting the checking experiment would not be considered as fault. Besides, the machine $M_r$ consists of the following special sequences.

Every sequence of $r$ distinct READ operations identifies the initial state of $M_r$ and constitute the distinguishing sequence ($D_0$). The sequence is given by $D_0 = R_0, R_1, \ldots, R_{r-1}$ where $R_0, R_1, \ldots, R_{r-1}$ respectively, are the $r$ distinct READ operations performed on each of the individual $r$ cells $0, 1, \ldots, r-1$ respectively. Further, the machine $M_r$ has a complete set of synchronizing sequences, since each state of $M_r$ can be changed to any other state by a sequence of at most $r$ WRITE operations performed on each of the individual cells of $M_r$. These sets of synchronizing sequences can be used suitably as the different transfer sequences for the concerned machine $M_r$.

Also, there is one interesting feature of the state diagram of $M_r$. The state diagram $G_r$ of $M_r$ is an *Eulerian Graph*. As, for example, the graph $G_2$ of the memory machine $M_2$ is obviously an Eulerian graph, because there are exactly $3r$ arcs (due to $r$ READ's and $2r$ WRITE's) entering every node (state) of $G_2$, while there are also $3r$ arcs leaving every state in the transition diagram as shown in Fig. 7.1. The subgraph $H_r$ of $G_r$ formed by removing all self loops also becomes Eulerian. The Eulerian graph $H_2$ of $G_2$ for the machine $M_2$ is shown in Fig. 7.2.

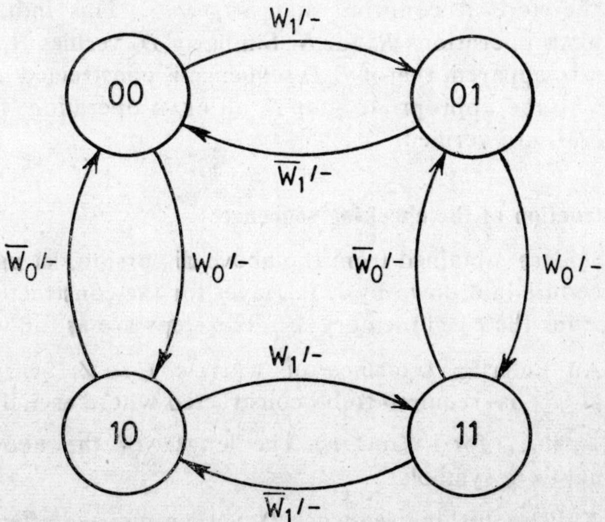


Fig. 7.2 Subgraph $H_2$ of the transition diagram $G_2$.

The verification of the correctness of WRITE operations is done assuming that the READ operations are fault free. It is evident from the Eulerian graph $H_r$, that for any machine $M_r$, any distinct write input changes the state of the machine $M_r$. A sequence $\lambda'$ of all distinct WRITE inputs which changes the state of $M_r$ from one particular state to another, corresponds to a path in $H_r$. An efficient method of checking all distinct WRITE operations is to insert the distinguishing sequence $D_0 = R_0R_1 \ldots R_{r-1}$ after each element which constitutes the sequence $\lambda'$. Similarly, each WRITE self-loop $\hat{W}_i$ can be verified by applying the sequence $\hat{W}_iD_0$ to $M_r$ when it is in the appropriate state.

For verifying $r$ READ transitions associated with every state a distinguishing sequence $D_i$ is obtained by cyclically shifting $D_0$, $i$ places to the left, i.e.,

$$D_i = R_iR_{i+1} \ldots R_{r-1}R_0 \ldots R_{i-2}R_{i-1}$$

and a sequence $D = D_0D_1D_2 \ldots D_{r-1}$ is formed by concatenation of each $D_i$ where $i = 0, 1, 2, \ldots, r-1$. Each READ transition can now be verified if a sequence $D_0D$ is applied to $M_r$ in every state. To make the statement clear consider the detail structure of $D_0D$ as follows:

$$\overbrace{\underset{t_0}{} R_0 \underset{t_1}{} R_1 \underset{t_2}{} \ldots R_{r-1}}^{D_0} \quad \overbrace{\underset{t_0'}{} R_0 \underset{t_1'}{} R_1 \underset{t_2'}{} \ldots R_{r-1}}^{D_0} \quad \overbrace{\underset{t_0''}{} R_1 \; R_2 \ldots R_0}^{D_1} \ldots D_{r-1}$$

Let $S(t)$ denote the state of the (fault free) memory at time instant $t$. If $M_r$ gives the correct response to $D_0D_0$ we conclude, $S(t_0) = S(t_0') =$ expected state $S$. Therefore, $D_0$ maps $S$ into $S$, and $S(t_0'') = S$. Now $D_1$ is a subsequence of $D_0D_0$ and maps $S(t_1)$ into $S(t_1')$. Since $D_1$ is applied immediately after $D_0D_0$, a valid response to this indicates that $S(t_0'') = S(t_1)$. But $S(t_0'') = S$, therefore it confirms that $S(t_1) = S$. This indicates that $D_1$ verifies the READ operations $R_0$ for $S$. Similarly $D_2$ verifies $R_1$ and so on. Therefore, it is apparent that if $D_0D$ sequence is constructed and applied subsequently to the appropriate state $S$, all READ operations for state $S$ of the machine $M_r$ are verified.

### 7.24 Construction of the checking sequence

With the ideas obtained from the above discussion, let us present the distinct procedure laid down by J. P. Hayes for the construction of checking sequence for the $r$ cell memory $M_r$. The steps are as follows:

*Step 1*: An Eulerian sequence of WRITE's $E = V_1V_2 \ldots V_{rq}$ for $H_r$ beginning at $S_0$ is required to be constructed where each $V_i$ belongs to $(\hat{W}_0, \hat{W}_1, \ldots, \hat{W}_{r-1})$ for $1 \leqslant i \leqslant rq$. The length of the above sequence becomes equal to $rq$ symbols.

*Step 2*: A distinguishing sequence $D_0$ is then inserted after each WRITE symbol in the above mentioned sequence $E$ to form a modified sequence

$T$ as follows:

$$T = V_1D_0V_2D_0 \ldots V_{rq}D_0$$

The length of the sequence $T$ obviously becomes equal to $(rq + r^2q)$ symbols.

*Step 3(a)*: For testing all the self-loops in $G_r$ corresponding to every state $S_i$, a sequence $L_i$ is formed as follows.

Let $S_i = (y_{i,0}, y_{i,1}, \ldots, y_{i,r-1})$ be any state of the memory $M_r$. Also let us take $\widehat{W}_j = W_j$ if $y_{i,j} = 1$ and $\widehat{W}_j = \overline{W}_j$ if $y_{i,j} = 0$, for $0 \leqslant j \leqslant r - 1$ and define $L_i = \widehat{W}_0D_0\widehat{W}_1D_0 \ldots \widehat{W}_{r-1}D_0D$ for $0 \leqslant i \leqslant q - 1$. The sequence $L_i$ would test all the self loops of $S_i$. The length of any $L_i$ sequence would be $(r + 2r^2)$ symbols and for all the $q$-states, i.e., for $0 \leqslant i \leqslant q - 1$, the length would be $(r + 2r^2)\ q$ symbols.

*Step 3(b)*: We have noticed in Step 2 that $T$ sequence causes the machine $M_r$ to go round through all the individual $q$ states of the machine. After each $V_jD_0$ in the sequence $T$, the machine $M_r$ resides in the different known initial state identified by the application of the distinguishing sequence $D_0$. Therefore, individual $L_i$ sequences formed earlier in above step 3(a) may now be inserted at the appropriate state after each $V_jD_0$ in the sequence $T$, to form the resulting sequence $T'$ for the machine. The resulting sequence $T'$ would have a length of $(2r + 3r^2)q$ symbols.

*Step 2*: The above sequence $T'$ forms the checking sequence of the memory if we assume the memory was set to the initial state at the beginning. Let $\lambda_0$ be a sequence that takes the machine $M_r$ from any state to the initial state $S_0$. Therefore, the sequence $T'$ is required to be preceded by a sequence $\lambda_0D_0$ only to obtain a general expression for the checking sequence $C_r$. Therefore, we have $C_r = \lambda_0D_0T'$.

EXAMPLE: Consider a Random access memory $M_2$ comprising of two cells only. The state table for the logical model of $M_2$ is shown in Table 7.1. The corresponding transition diagram $G_2$ and the subgraph $H_2$ are shown in Fig. 7.1 and Fig. 7.2 respectively.

Let $S_0 = (0, 0)$ be the initial state of the machine $M_2$. The steps for the designing of checking experiment would be as follows:

*Step 1*: Construct a sequence $E$ by tracing the Eulerian path through the subgraph $H_2$ by starting and finishing with the state $S_0$ as follows:

$$E = W_1W_0\overline{W}_1\overline{W}_0W_0W_1\overline{W}_0\overline{W}_1$$

*Step 2*: Insert the distinguishing sequence $D_0 = R_0R_1$ after each WRITE symbol in the above sequence $E$, to form the sequence $T$ as follows:

$$T = \underset{t_0}{W_1D_0}\ \underset{t_1}{W_0D_0}\ \underset{t_2}{\overline{W}_1D_0}\ \underset{t_3}{\overline{W}_0D_0}W_0D_0W_1D_0\overline{W}_0D_0\overline{W}_1D_0$$

*Step 3*: For checking the self loops in the graph $G_2$ for every state, the following four self loop sequence $L_0$, $L_1$, $L_2$ and $L_3$ are formed as follows.

Since the distinguishing sequence $D_0 = R_0R_1$, the sequence $D$ would be

$D = R_0R_1R_1R_0$ and the four self-loop test sequences are represented as follows:

$$L_0 = \overline{W}_0D_0\overline{W}_1D_0D$$

$$L_1 = \overline{W}_0D_0W_1D_0D$$

$$L_2 = W_0D_0\overline{W}_1D_0D$$

$$L_3 = W_0D_0W_1D_0D$$

Now if the sequence $T$ as formed in the Step 2 is applied to the memory/(machine $M_2$), it should be in the state $S_0$, $S_1S_3$, $S_2$ at the time instants $t_0$, $t_1$, $t_2$, $t_3$ respectively as indicated in the Step 2 in the sequence $T$. Therefore, it is required now to insert $L_0$, $L_1$, $L_2$ and $L_3$ at the appropriate places in the sequence $T$ to obtain the modified sequence $T'$. Thus $T'$ would be represented as

$$T' = L_0W_1D_0L_1W_0D_0L_3\overline{W}_1D_0L_2\overline{W}_0D_0W_0D_0W_1D_0\overline{W}_0D_0\overline{W}_1D_0$$

*Step 4*: Next, we precede the sequence $T'$ by the initializing sequence $\lambda_0D_0$ where $\lambda_0 = \overline{W}_0\overline{W}_1$, to obtain the preset checking sequence $C_2$ for the machine $M_2$ of the two cells random access memory as follows:

$$C_2 = \lambda_0D_0L_0W_1D_0L_1W_0D_0L_3\overline{W}_1D_0L_2\overline{W}_0D_0W_0D_0W_1D_0\overline{W}_0D_0\overline{W}_1D_0$$

The length of the above checking sequence is 68 symbols for the two cells memory only.

The length of the checking sequence thus obtained would become exponential in $r$, the number of cells in the memory. Therefore, the designing of the checking sequence does not really become computationally feasible. For obtaining the reasonable test sequences, Hayes [104] had suggested two approaches whereby some restriction may be imposed to the pattern sensitive faults.

One approach is to restrict the error behaviour associated with a particular cell $C_i$ to a well defined neighbourhood $N_i$ of the said cell $C_i$. It seems to have considerable practical justification, though it may not always be easy to specify definitely the neighbourhoods. When the cell $C_i$ is considered to be a member in the neighbourhood $N_i$, the other members of $N_i$ may be termed as adjacent to the cell $C_i$. The adjacency among the cells in a certain neighbourhood may be considered in several ways: (i) The addresses of the cells in $N_i$ may have certain well defined relationship with one another. Interconnections among the cells in the neighbourhood may lead to hazard or race in addressing circuitry whereby a cell $C_j$ whose binary address $j$ differs from that of cell $C_i$ may get addressed during the READ/WRITE operation addressed to the cell $C_i$. (ii) The input/output leads of those members of $N_i$ that are physically adjacent only may result in electromagnetic interference.

Considering the adjacency behaviour among the cells as mentioned above, the entire memory $M_r$ may be thought to be consisting of a set of neighbourhoods $N = (N_0, N_1, \ldots, N_{r-1})$. A local pattern sensitive faults (PSF) with respect to $N$ is defined to be any PSF $F(N)=(W_i^F \overline{W}_i^F, R_i^F, g^F)$ with the following two properties, viz., (i) functions $(X_i^F, g^F)$ where $X_i^F$ denotes $W_i^F$, $\overline{W}_i^F$, or $R_i^F$ are independent of the contents of all cells $C_j \notin N_i$, and (ii) $X_i^F$ cannot alter the content of $C_j \notin N_i$. Thus, considering the local PSF only, we are permitted to restrict our attention to a well-defined neighbourhood of cells only and may not enlarge our attention to take into consideration of all the $r$ number of cells in the entire memory, while constructing the checking experiment of the said memory.

Alternatively, the reduction in the length of the checking sequence would be affected by restricting the kinds of faults that can occur. Unlike unrestricted faults, here one would restrict the nature and the scope of interaction of a cell with its neighbouring cells. For example, operation WRITE 1 into cell $C_i$ may more likely to write 1 into another cell $C_j$ then it is to write 0 into the cell $C_j$. Therefore, it may be possible to postulate a class of monotonic WRITE faults, where $W_i$ (or $\overline{W}_i$) would only increase the number of 1's (or 0's) in $N_i$. The restriction that would be introduced on the nature of the faults in this manner would no doubt reduce the length of the checking sequence, but the designing of the checking experiment would simultaneously become further complicated.

### *7.3 Bit-sliced microprocessor

Among the various classes of LSI circuit components available these days, bit-sliced microprocessors are very important in the sense that they can be interconnected in a regular fashion for constructing more useful types of digital systems. A bit-sliced microprocessor is an array of $n$-identical ICs called slices, each of which is a simple processor of operands of length $k$ bits, where $k$ is typically 2 or 4. Here interconnections between $n$ slices are constructed in such a way that the entire array becomes a processor of $nk$ bit operands. Bit-slices provide structural regularity in system design which facilitates fault diagnosis and test generations.

T. Sridhar and J. P. Hayes [211] have recently developed an analytical technique for test generations of bit-sliced microprocessors. For this purpose, they developed a formal model for a bit-sliced microprocessor which had all the main features of the commercially available microprocessor like AMD 2901 of Advanced Micro Devices. Figure 7.3 shows the circuit model of the 1-bit processor cell considered for this purpose with its micro-instructions and control fields. For arriving at reasonably realistic fault analysis, the operand size of the cell is taken to be one bit only. Of course, the one-bit cell differs from the commercially available processor slices of these days. This is due to the fact that, in reality, no bit-sliced processors having one bit word length are being manufactured, although one bit non-bit-sliced microprocessors such as Motorola 14500

*Illustrations (tables and figures etc.) in Section 7.3 are from the paper referred [211] in the bibliography and are presented with the permission of IEEE (copyright © 1980 IEEE).

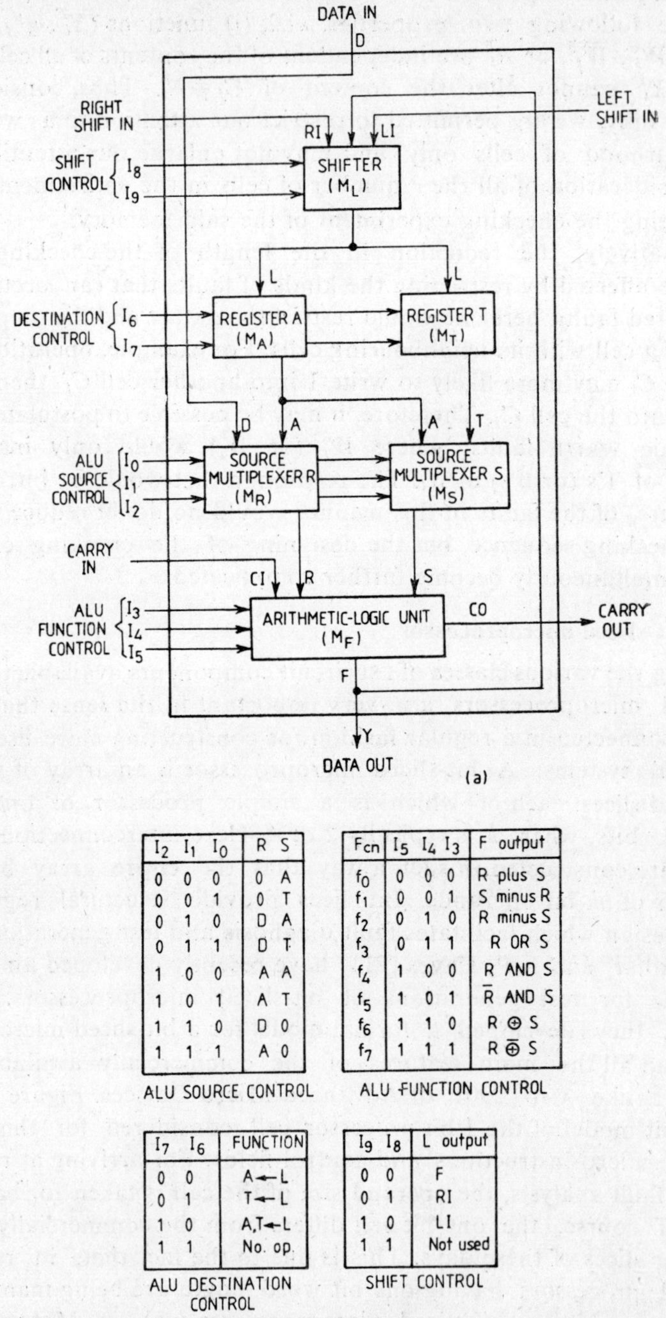


| $I_2$ | $I_1$ | $I_0$ | R | S |
|---|---|---|---|---|
| 0 | 0 | 0 | 0 | A |
| 0 | 0 | 1 | 0 | T |
| 0 | 1 | 0 | D | A |
| 0 | 1 | 1 | D | T |
| 1 | 0 | 0 | A | A |
| 1 | 0 | 1 | A | T |
| 1 | 1 | 0 | D | 0 |
| 1 | 1 | 1 | A | 0 |

ALU SOURCE CONTROL

| $F_{cn}$ | $I_5$ | $I_4$ | $I_3$ | F output |
|---|---|---|---|---|
| $f_0$ | 0 | 0 | 0 | R plus S |
| $f_1$ | 0 | 0 | 1 | S minus R |
| $f_2$ | 0 | 1 | 0 | R minus S |
| $f_3$ | 0 | 1 | 1 | R OR S |
| $f_4$ | 1 | 0 | 0 | R AND S |
| $f_5$ | 1 | 0 | 1 | $\bar{R}$ AND S |
| $f_6$ | 1 | 1 | 0 | $R \oplus S$ |
| $f_7$ | 1 | 1 | 1 | $\overline{R \oplus S}$ |

ALU FUNCTION CONTROL

| $I_7$ | $I_6$ | FUNCTION |
|---|---|---|
| 0 | 0 | A←L |
| 0 | 1 | T←L |
| 1 | 0 | A,T←L |
| 1 | 1 | No. op. |

ALU DESTINATION CONTROL

| $I_9$ | $I_8$ | L output |
|---|---|---|
| 0 | 0 | F |
| 0 | 1 | RI |
| 1 | 0 | LI |
| 1 | 1 | Not used |

SHIFT CONTROL

(b)

Fig. 7.3 (a) 1-bit processor model $U$, (b) its microinstruction control fields.

are not uncommon. The overall structure of the cell is very much similar to that of AMD 2901 processor and also it resembles the processor 74S481 as it also contains two working registers A and T only similar to that of 74S481 processor cell. While forming the array structures of these cells, the carry-look-ahead circuit is not taken into consideration instead, the ripple carry propagation which facilitates the formation of iterative logic array (ILA) structure, is only thought to be present. Because of the ILA formation the test developed for a single cell $U$ can easily be extended to the array of slices.

As shown in Fig. 7.3, the 1-bit processor cell consists of six basic modules. There are two scratch pad registers viz. an accumulator $M_A$ and one additional temporary register $M_T$, two multiplexer modules $M_R$ and $M_S$, one arithmetic and logic unit (ALU) module $M_F$ and one shifter module $M_L$. The internal structure of any of these modules is unimportant in the sense that such information is normally not available to the system designer. A fault model based on the functions performed by each of these modules is defined with a restriction that not more than one module in the cell can become faulty at any time.

Let $M$ be any combinational or synchronous sequential circuit module embedded inside the cell $U$. Let $O$ denote the function realised by $M$ and let $S$ be the number of internal states of $M$ such that if $S = 1$, the $M$ is known to be a combinational circuit module and if $S > 1$, the module is considered to be a synchronous sequential circuit module. A malfunction $F$ of $M$ is known as a functional fault of $M$, if due to the fault $F$, the module $M$ changes to a new module $M^F$ realizing $O^F$, where $O \neq O^F$, and the number of states $S^F$ of $M^F$ do not exceed the number of states $S$ of $M$.

### 7.3.1 *Test generation for a single cell U*

A general approach laid down by Sridhar and Hayes [211] for the test generation of a bit-sliced microprocessor cell, consisting of $k$ modules $M_1, M_2, \ldots, M_k$ will be presented now. Let $T_i$ be a test that detects all functional fault in an isolated module $M_i$. If $M_i$ happens to be a combinational circuit module, then for $T_i$ one would have to consider all the $2^n$ inputs to the module, where $n$ is the number of (primary) input lines of $M_i$. On the other hand, in case $M_i$ happens to be synchronous sequential circuit, then the checking experiment can be developed for generating test sequence $T_i$. As $M_i$ is an embedded internal component of the unit under test, i.e., the cell $U$, the faults in $M_i$ are to be detected by applying a test sequence $T_i^*$ at the accessable external input terminals of the cell $U$. This should ensure that $T_i$ is applied internally to the component $M_i$ and additionally ensure that the responses of $M_i$ may be propagated to the observable output terminals of the cell $U$. A composite test sequence $T_u$ for the entire cell $U$ is obtained by concatenating the various $T_i^*$ test sequences, viz., $T_u = T_1^* T_2^* \ldots T_k^*$. As on principle, only one module may become faulty at any time, the test sets for the several modules may sometimes be merged properly to reduce the number of test patterns for the cell $U$ as a whole.

The basic cell $U$ contains six modules as discussed earlier, and out of these, the modules $M_L$, $M_R$, $M_S$ and $M_F$ are combinational circuit module, while $M_A$ and $M_T$ are synchronous sequential module. Each of these modules has one or two output lines. The output signal of each of these modules must be observed at one of the output lines $Y$ or $CO$ of the cell $U$. As an example, let us enumerate the testing procedure for the shifter module $M_L$. The module $M_L$ has five input lines consisting of $I_8$, $I_9$, $RI$, $F$, $LI$ only, and the test pattern $T_L$ would consist of all $2^5 = 32$ possible combinations of these input variables. The responses of the module $M_L$ can be routed to the observable output terminal $Y$ as follows. Referring to Figure 7.4 the output signal $L$ of $M_L$ is sent through the register $M_A$

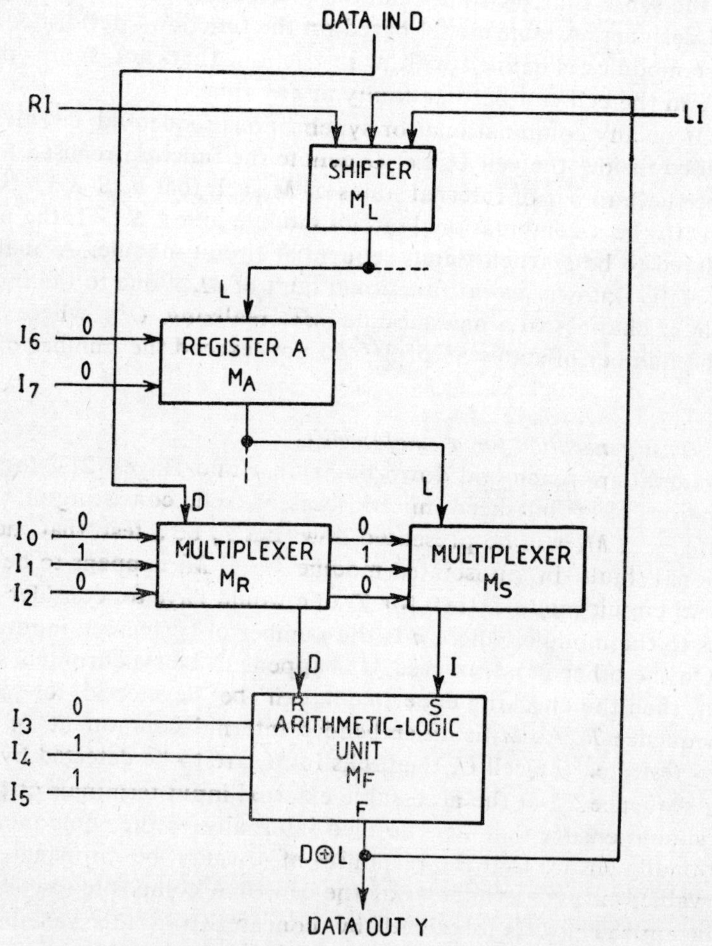


Fig. 7.4 A set-up of 1-bit processor module $U$ for testing the shifter module $M_L$.

by setting the control signals $I_7$ and $I_6$ both to 0. It is then propagated to the right input port $S$ of the ALU module $M_F$ by setting $I_0$, $I_1$, $I_2$ as 0, 1, 0 respectively. This also ensures that the input data line $D$ is applied to the left input port $R$ of $M_F$. Further the control signals $I_3$, $I_4$, $I_5$ and CI are set to 0, 1, 1 and $d$ (don't care) value respectively to specify the exclusive-OR function which results in the ALU output signal $F = D \oplus L$. As $F$ is connected here to the primary output line $Y$ of the cell and $D$ is a primary input line of the cell, the output signal $L$ of the shifter module $M_L$ has thus been made available at the output line $Y$ of the cell $U$. Besides this, the next value of $F$ that is to be applied as an input signal to $M_L$ can also be controlled externally by the appropriate selection of the value for $D$. Thus $T_L^*$ can be constructed easily to have the same number of test patterns as $T_L$ which requircs 32 test patterns only. Like $M_L$ the ALU module $M_F$ may also be tested which requires 64 test patterns and also similarly the multiplexers $M_R$ and $M_S$ require 32 tests, most of which can be combined together with that of the tests for the ALU and may be merged properly. It was shown by Sridhar and Hayes that at most 98 test patterns would become enough for testing all the four combinational circuit modules of the cell $U$.

The register modules $M_A$ and $M_T$ are assumed to be Moore type sequential machines each with a state table and state diagram as shown in Tables 7.2 and 7.3, and Figures 7.5 and 7.6 respectively.

Table 7.2 Transition table of the register module $M_A$

| Present state | Next states for inputs $I_7I_6L$ | | | | | | | | Present output |
|---|---|---|---|---|---|---|---|---|---|
| | 000 | 010 | 100 | 110 | 001 | 011 | 101 | 111 | |
| 0 | 0 | 0 | 0 | 0 | 1 | 0 | 1 | 0 | 0 |
| 1 | 0 | 1 | 0 | 1 | 1 | 1 | 1 | 1 | 1 |

Table 7.3 Transition table of the register module $M_T$

| Present state | Next states for inputs $I_7I_6L$ | | | | | | | | Present output |
|---|---|---|---|---|---|---|---|---|---|
| | 000 | 010 | 100 | 110 | 001 | 011 | 101 | 111 | |
| 0 | 0 | 0 | 0 | 0 | 0 | 1 | 1 | 0 | 0 |
| 1 | 1 | 0 | 0 | 1 | 1 | 1 | 1 | 1 | 1 |

Evidently the state diagram is Eulerian, i.e., there is a single path passing through every transition of the state diagram exactly once. An Eulerian path which passes through every transition exactly once corresponds to a minimum length checking sequence which detects all functional faults in the sequential module. This is because of the fact that the next state, which also registers the output signal, can be made available at the primary output terminal $Y$ of the cell $U$ by suitable selection of ALU

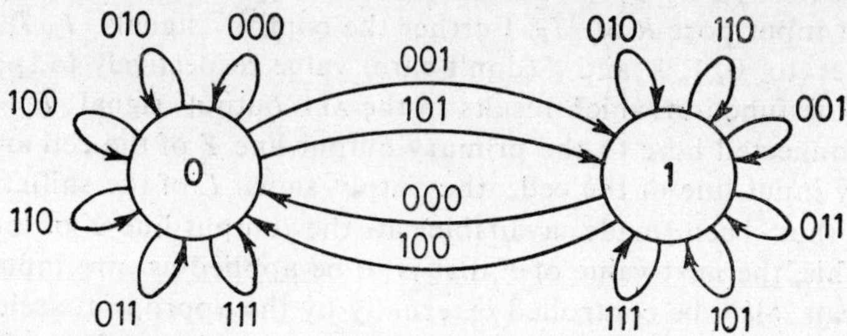


Fig. 7.5 Transition diagram for register $M_A$ with inputs $I_7 I_6 L$.

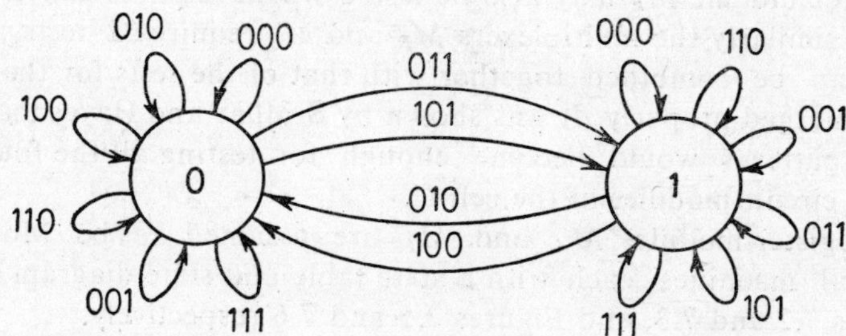


Fig. 7.6 Transition diagram for register $M_T$ with inputs $I_7 I_6 L$.

source control and function control signals as was done for the shifter module $M_L$ earlier. Thus an optimal checking sequence for $M_A$ and $M_T$ can easily be derived and is shown below:

| | | | | | | | | | | |
|---|---|---|---|---|---|---|---|---|---|---|
| Inputs of $M_A$ and $M_T$ $I_7$: | | 0 | 0 | 1 | 1 | 1 | 0 | 1 | 1 | 1 |
| $I_6$: | | 0 | 1 | 1 | 1 | 0 | 1 | 0 | 1 | 1 |
| $L$: | | 0 | 0 | 0 | 1 | 1 | 1 | 1 | 0 | 1 |
| Output (state) of $M_A$: | 0 | 0 | 0 | 0 | 0 | 1 | 1 | 1 | 1 | 1 |
| Output (state) of $M_T$: | 0 | 0 | 0 | 0 | 0 | 1 | 1 | 1 | 1 | 1 |

| | | | | | | | |
|---|---|---|---|---|---|---|---|
| Inputs of $M_A$ and $M_T$ $I_7$: | 0 | 0 | 0 | 0 | 1 | 0 | 1 |
| $I_6$: | 0 | 0 | 1 | 0 | 0 | 1 | 0 |
| $L$: | 0 | 1 | 0 | 1 | 0 | 1 | 0 |
| Output (state) of $M_A$: | 0 | 1 | 1 | 1 | 0 | 0 | 0 |
| Output (state) of $M_T$: | 1 | 1 | 0 | 0 | 0 | 1 | 0 |

The input $L$ signal required for the next test input in the above sequence is obtained by suitably selecting the source control and ALU function to indicate $L = F = A(T)$ or $L = F = \bar{A}(\bar{T})$ for the two registers. This also enables the next state to be observed at the $Y$ output end. In this way, optimal test sequences $T_A$ and $T_T$ are easily constructed and as the test sequences $T_A$ and $T_T$ are essentially the same, they can be merged into a single sequence $T_{AT}$ of length 16 which tests the sequential modules of the cell $U$. Thus the entire processor $U$ can be tested by a sequence $T$ containing $98 + 16 = 114$ test patterns. This is even less than twice the minimum number of test patterns (64) to test the ALU module $M_F$ alone.

### 7.3.2 *Testing approach for the k-bit version of the cell U*

The 1-bit slice $U$ of Figure 7.3 can be extended to form $U^k$, the $k$-bit version of the original 1-bit slice $U$. The $k$-bit processor slice is shown in Figure 7.7. The replication technique is utilized for expanding the word

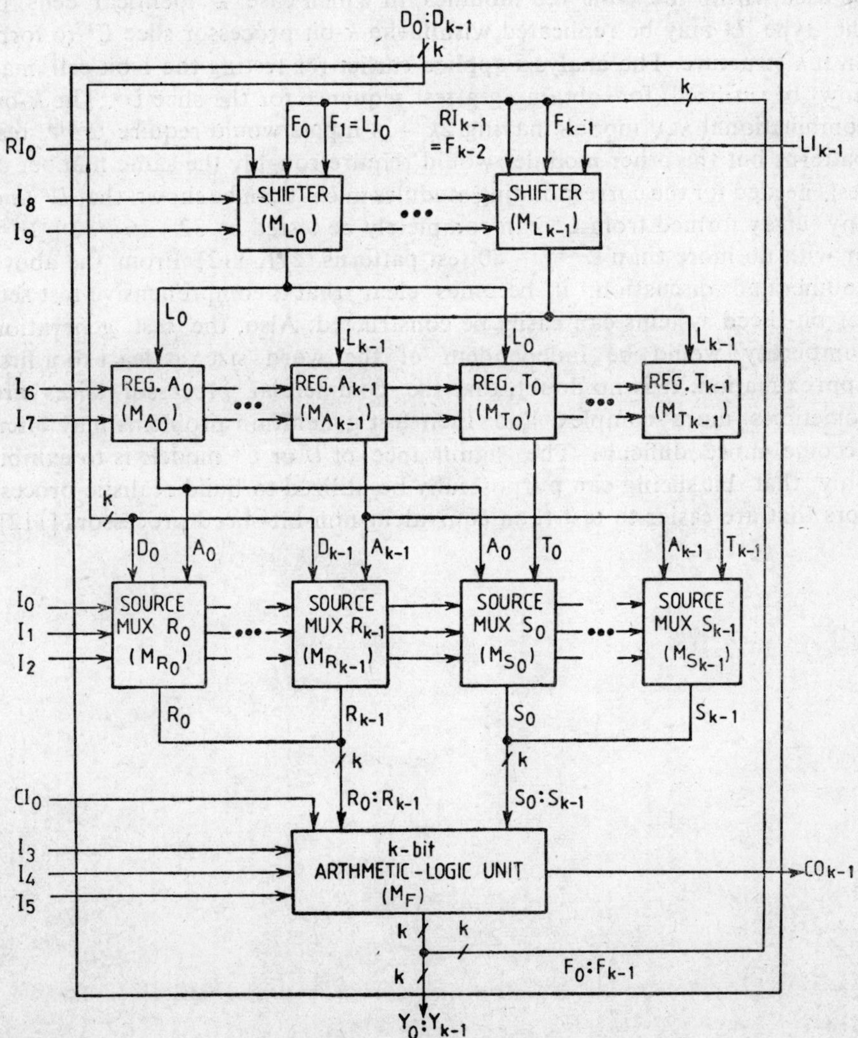


Fig. 7.7 $k$-bit version $U^k$ of cell $U$.

size of the processor from 1 to $k$. This can be understood by considering the multiplexer module $M_R$. In $U^k$ the $M_R$ is replicated $k$-times to form the $k$ multiplexer modules $M_{R0}, M_{R1}, \ldots, M_{Rk-1} = M_{R0} : M_{Rk-1}$. The module $M_{Ri}$, $0 \leqslant i \leqslant k-1$, operates on the $i$th bit of the $k$-bit word obtained from the direct input bus $D_0 : D_{k-1}$, and the corresponding $i$th bit obtained from the registers $M_{A0} : M_{Ak-1}$. The modules $M_L$, $M_A$, $M_T$ and $M_S$ of the cell $U$ are also being expanded in the cell $U^k$ in the same manner. In the ALU module $M_F$, the above expansion techniques are not being utilized and a single ALU module operating on the $k$-bit operands $R_0 : R_{k-1}$ and $S_0 : S_{k-1}$ are being retained while forming the $k$-bit version. This requires the use of carry-look-ahead scheme within $U^k$. If such speed up techniques are not to be utilized, then ripple carry propagation may be used within the $k$-bit ALU modules, in which case $k$ identical cells of the type $U$ may be replicated within the $k$-bit processor slice $U^k$ to form an ILA structure. The analysis applied earlier for testing the 1-bit cell may now be utilized for obtaining a test sequence for the slice $U^k$. The $k$-bit combinational ALU module having $2k + 4$ inputs would require $2^{2k+4}$ test patterns but the other modules would require roughly the same number of tests needed for the corresponding modules in $U$. It can be shown that $U^k$ and any array formed from $U^k$ can completely be tested by $32+16+32+2^{2k+4}$ or with no more than $2^{2k+4} + 80$ test patterns [211, 112]. From the above enumerated discussion, it becomes clear that a comprehensive test sets for bit-sliced systems can easily be constructed. Also, the test generation complexity would be independent of the word size at least to a first approximation. It is no doubt, that the commercial processor slices are sometimes more complex and their test generation problems may often become more difficult. The significance of $U$ or $U^k$ models is to exhibit only that bit slicing can purposefully be utilized to build realistic processors that are easier to test than equivalent non-bit-sliced processors [112].

Appendix

# Microprocessors/Microprocessor-Based Design Testing

Today we are living in the era of Large Scale Integration (LSI) technology that helps making digital system in encapsulated form. If LSI of today is measured and represented by a term called 1 00 000 transistors on a chip, tomorrow it would possibly be increased by an order of higher magnitude of 1 000 000 transistors on the same area and it may yet further be increased in the years to come. As the packaging density increases, the use of LSI components is frequently being made in system design, which leaves fewer test points available and the data streams at the available test points are becoming complex for any purpose of fault detection or diagnosis. But for testing any system constructed with the LSI devices, one would have to apply some suitable stimulus to the circuit and have to analyze the resulting data patterns to locate the faulty component so that it may be replaced by another field replaceable component, bringing the circuit back to the operation again.

Looking back at the historical development of integrated circuits, single-gate functions (AND, OR, NAND, NOR) were first developed. Truth-table testing of these devices requires anyway from two test patterns for an inverter to sixteen for a four-input gate. There is no difficulty of manually-generating such patterns. But when more bigger circuit, say an arithmetic logic unit (ALU) has been prepared in a chip, the manual generation of the test pattern for the same becomes extremely difficult. Here comes the usage of computer simulation and routine analysis for generating the thousands of tests patterns required for testing the concerned chip. Testing becomes further complex when it is applied to, say, a calculator chip. A calculator chip can store an answer to be used in some later calculations. The outputs for a calculator not only depend on the present inputs but also on the preceding sequence of inputs, thus resulting in the 'sequential' or 'random logic' behaviour of the system. Even utilizing the technique of computer simulation, the number of test patterns and test cycles required for proper testing of the said chip becomes very high.

For testing integrated circuits in encapsulated form, the test patterns are either produced manually by a design or test engineer, or automatically, by special hardware or software implemented algorithms called test generation program. The set of test patterns with the correct responses is known as *fault dictionary* and the testing is implemented sometimes by sorting such test data, and is known as *stored response testing*.

With the advent of microprocessor, the current trend of designing digital systems changed drastically. Now-a-days, the entire digital system most often takes the form of a microcomputer comprising of a microprocessor (a programmable processor consisting of a small number of integrated circuits or often just a single IC) which acts as central processing unit or system controller, read only memories (ROM), random access memories (RAM) and input/output circuits. Microprocessor-based systems are difficult to test because of the following reasons, namely, (i) Like any other LSI, it also contains thousands of basic components (gates) and interconnecting lines, all are individually subject to failure; (ii) Access to internal components and lines is severely limited because of the limited number of external accessible input/output connections that are available; (iii) A large number of test patterns requires to be simulated because of the large possibility of faults; (iv) The complex failure modes like pattern-sensitivity may occur; (v) The system designers are generally being provided with only those specifications which include a register-level block diagrams, a listing of microprocessor's instruction set and some information regarding the system's timing only.

A classical approach of testing a microprocessor is developed by treating it as a black box with some limited instruction set. The problem here becomes that of easily producing the test patterns for different sequences of instructions or data as different sensitivities are discovered. This is being carried out generally by the following two approaches. The first approach requires availability of a known good device known as 'gold device'. An interactive software routine is required and the operator provides the instruction sequence and input data to the software routine. He then applies these inputs to the gold device and reads back the assumed correct outputs. These outputs along with the inputs are then stored on some bulk storing medium (say a disk), for later testing of unknown devices. Therefore, the type of testing is nothing but a *comparison testing* and the gold device acts as a reference against which the other is compared. In the second approach, the interactive software and the gold device are being replaced by a software simulator for the microprocessor. In this case, the device outputs are being generated under software control. However, either of these approaches has the limitation as it tests only a minute fraction of the possible combination of instructions and data that the device might encounter in actual practice.

Besides this classical approach, two new techniques can provide comprehensive tests of microprocessors and these are known as *modular sensorialization* and *algorithmic pattern generation*. These may be used individually or both together, but eventually simplify the testing of microprocessors and reduce the amount of memory space needed for storing the test data and the time required for producing and applying the stimuli to the unit under test (UUT). A modular sensorialization approach is being found effective where the device has been thought to be consisting of different functional subsystems or modules. In microprocessor-based system, because the bus type of organization is used, functional modules in the system can be tested separately in an independent fashion. The bus-structured architecture allows access to critical buses which go to many different modules on the computer board. From Figure A.1, it is clear that the data bus is involved among the microprocessor module, the ROM module, the RAM module, and the I/O controller module. By applying an

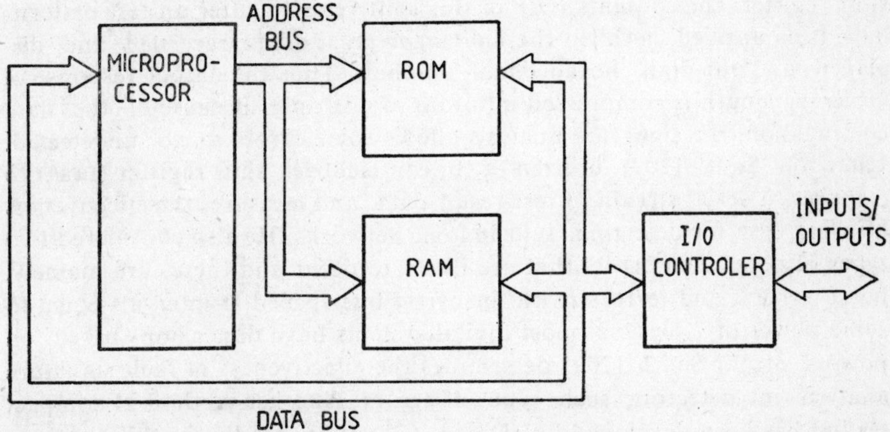


Fig. A.1 Bus structured microcomputer.

external excess to the data bus, three of the four modules can be turned off the data bus, i.e., their outputs (may be tristated) can be put to high impedance. By doing this, the data bus could be used to drive the fourth module, as if it carries the primary input (or primary output) of the said module. Similarly, by exercizing external control, the address bus can be made useful in controlling the test patterns to the microcomputer board. These buses, therefore, practically partition the microcomputer board uniquely, so that the testing of subunits can be accomplished nicely. A drawback of bus-structured designs results from the faults associated with the bus itself. If a bus is stuck high/low, any module or the bus trace itself may be faulty. Normally, the testing is done by finding the location of a fault from the voltage information. To isolate a failure for the bus only, we may require current measurements, which are much difficult to be carried out.

Algorithmic pattern generation is applied generally, after the sequences that are appropriate for testing a microprocessor (or other digital IC) have been devised, may be through the approaches of modular sensorialization, or some other method. Mostly in automatic test systems, the sequence of bit patterns necessary to simulate a UUT is generated by a computer program. During the test, these patterns are transferred into a large random access memory area and then allowed to be transferred in a burst to the UUT. If $S$ is the input sequence (or stimulus) applied to a logic circuit $N$ and $Z$ is its corresponding output response, then the observed result or the *signature* of the test may be as long as $Z$ or a *compact function* of $Z$. A compact testing method known as *transition counting* computes a number of logical transitions (a 0 changing to 1 and *vice versa*) occurring in the output response at the test point [105]. Another compact testing scheme known as *signature analysis* is proposed by Hewlett-Packard Company [83] and is intended specially for microprocessor-based systems. The output response from the UUT is passed through a 16-bit feed back shift register; the contents $f(R)$ of this shift register, after all test patterns have been applied, is called the *fault signature*, and is recorded and displayed as four-digit hexadecimal number. Thus the output response of arbitrary length is compressed into four digits only. Because of the data compression, the signature analysis allows some errors to go undetected. Later on Smith [209] utilized a linear feedback shift register (LFSR) to compress a serial stream of test result data, and measures the effectiveness of using LFSR for detecting faults in logic networks. He also considered two types of dependent errors that are likely to occur and these are namely, burst errors and errors with incorrect bits spaced at intervals equal to some power of 2, because most digital systems have dimensions based on powers of 2. Smith [209] determines the effectiveness of fault signature analysis in detecting such types of errors. Another method of compact testing has been developed by David [67] using feedback shift register (FSR) is shown to have the maximum resolution and maximum distinction potential (a new notion introduced by David himself) that can be found for an $m$-bit signature.

Of late, since the programmed testing has become important in testing microprocessor-based system, let us discuss the issue in more details. The microprocessor is program-controlled, and the program can apply the test patterns to the UUT so that appropriate test patterns may be applied to the major register level modules, all of which must have to be accessible with different instruction set. In practice, a representative set of input patterns are applied to the UUT causing these patterns to go through a representative set of state transitions and these representative pattern sets are constructed on heuristic considerations only. A test program for a microprocessor is built-up in certain steps, each testing a related group of instructions or components. If a group has been found to be fault free, then it may be used to test the other groups. Selection and sequencing of these steps are complicated because there exists a considerable overlap among the components affected by different instructions.

The method of Chiang and McCaskill [45] for constructing the test program for Intel 8080 (a popular 8-bit microprocessor from the Intel Corporation) is considered. The architecture of the Intel 8080 microprocessor is shown in Figure A.2. The 8080 contains a simple arithmetic-logic unit and six 8-bit general purpose registers, the latter may be paired to form three 16-bit registers. The address size of the memory is

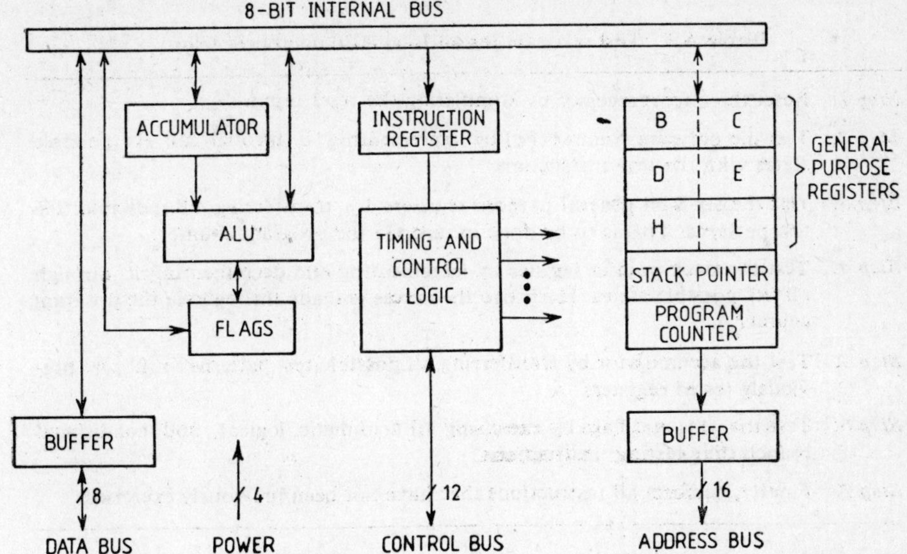


Fig. A.2 Architecture of Intel 8080 microprocessor.

of 16-bits; and the program counter and stack pointer registers are also of 16-bits. The main steps in an 8080 test program is shown in Table A.1. It is assumed that the system consisting of the microprocessor has been connected to an external tester which has access to its three buses, viz., the data, address and control buses. At first, the external tester resets the 8080-based UUT. Then it increments the 16-bit program counter (PC) through all its 65,536 states. This can be done by placing a single instruction NOP (no operation) on data (input) lines and causing the microprocessor to execute the instruction repeatedly. While the PC is being incremented, it automatically places its contents on the outgoing address bus lines where the tester would be able to observe and check the same. The checking would be accomplished rapidly by comparing the PC state to that of a hardware or software counter in the tester which also gets incremented on-line in step with the PC.

After doing this, the various general-purpose registers are checked by transferring 8-bit test patterns to them and checking the result simultaneously. All possible 256 test patterns are applied and the test patterns are generated algorithmically. Several data transfer operations like MOV, LXI, PCHL etc. are required for the implementation of the same; and these data transfer operations are to be checked individually earlier. After a pattern is applied to a register *R*, the tester is capable of inspecting the contents of *R* by transferring them to the PC via the high-level register HL; and as the PC has already been tested at the beginning, its contents are taken to be correct and are observed directly via the address bus. The remaining steps of the test program exercise the other components and instructions of the microprocessor in a similar way [110].

Table A.1 Test program for an Intel 8080 microprocessor

| | |
|---|---|
| *Step 1*: | Reset the microprocessor by stimulating the reset input. |
| *Step 2*: | Test the program counter (PC) by incrementing it through all its possible states with the NOP instruction. |
| *Step 3*: | Test the six 8-bit general purpose registers by transferring all possible 256 test patterns. This is to be done in turn via the program counter. |
| *Step 4*: | Test the stack pointer register by incrementing and decrementing it through all its possible states. Here also the access is made through via the program counter. |
| *Step 5*: | Test the accumulator by transferring all possible test patterns to it via previously tested registers. |
| *Step 6*: | Test the ALU and flags by exercising all arithmetic, logical, and conditional branch (flag testing) instructions. |
| *Step 7*: | Finally, perform all instructions that have not been previously exercised. |

Next, consider the testing of the entire microprocessor-based system with a typical bus-oriented architecture as shown in figure A.1. Again we have to take the help of an external tester that has access to the various system buses. Also, it is assumed that the external tester is capable of disconnecting parts of the system from the buses during testing; and this can be done either electrically or mechanically. It is to be noted here that input/output device testing may not be considered because it varies from device to device.

At first, it is to be seen that whether the PC inside the microprocessor is operating correctly. This can be done by using NOP operation as mentioned earlier but it is very much essential to isolate the microprocessor from the data bus during the test so that the external tester would be able to apply the instructions needed to increment the PC. In the free-running operation of the PC, the external tester monitors and checks the signals appearing on each of the system's address lines, may be through a compact testing method like signature analysis. Next, keeping the microprocessor in the free-running mode, the system ROMs would be tested. While doing this, the RAMs are to be disconnected from the data bus. Microprocessor could easily generate all memory addresses, therefore every ROM location would be accessed automatically and the tester would monitor the signals as they would appear on the data bus. Since the ROM contents are always fixed, a fixed signature can easily be associated with ROM. The remaining part of the system would then be tested through specific exercising programs, which are stored generally in the external tester or in the UUT's ROMs. The RAMs may be tested by programs required for

*Galpat testing*, though the widely used Galpat testing method would require about $4n^2$ test patterns to check an $n$-bit RAM [14, 110]. The input/output interface circuits normally resemble memory devices; therefore these may also be tested by memory-oriented check programs that have been developed recently. This is done by a technique called *look-back* and the output ports would be jumper connected to the input ports; and the processor would send the pattern to an output port and would check the same by reading back via the input port [110].

Thus, the testing of digital ICs or system constructed with digital ICs (it may be even microprocessor or microprocessor-based system) has become so vast and complicated that there is a need now-a-days to construct and fabricate integrated circuits with built-in efficient fault-detection capabilities [141]. McCluskey [110] has defined two terms called the '*controllability*' and '*observability*' as guidelines to the design engineers for constructing ICs, so that easy testing of the same may be achieved. Controllability may be defined as the measure of effectiveness with which test input patterns can be applied to the inputs of a subcircuit, $S(i)$, by exercising the primary inputs of the circuit $S$. Similarly, the observability is defined as the measure of effectiveness with which the responses of subcircuit $S(i)$ can be determined by observing the primary outputs of $S$. Therefore it is the motto of the designer of these days to design ICs by imposing better controllability and observability in it. An important as well as systematic design procedure for increasing the testability of LSI devices including microprocessors is to implement the scan-in/scan-out technique suggested by Williams and Angell [226] and is discussed earlier in section 4.7. A similar technique known as level sensitive scan design (LSSD) has been developed by IBM to incorporate testability in their structural design [158, 227].

# Bibliography


1. Abramovici, M. and Breuer, M. A., "On redundancy and fault detection in sequential circuits", *IEEE Trans. Comput.*, Vol. C–28, No. 11, pp. 864–865, Nov. 1979.

2. ——, "Multiple fault diagnosis in combinational circuits based on an effect-cause analysis", *IEEE Trans. Comput.*, Vol. C–29, No. 6, pp. 451–460, June 1980.

3. Abramovici, M., "A hierarchial, path-oriented approach to fault diagnosis in modular combinational circuits", *IEEE Trans. Comput.*, Vol. C–31, No. 7, pp. 672–677, July 1982.

4. Akers, S. B., "Universal test sets for logic networks", *IEEE Trans. Comput.*, Vol. C–22, No. 9, pp. 835–839, Sept. 1973.

5. —— "Fault diagnosis as a graph coloring problem", *IEEE Trans. Comput.*, Vol. C–23, No. 7, pp. 706–713, July 1974.

6. ——, "Binary decision diagram", *IEEE Trans. Comput.* Vol. C–27, No. 6, pp. 509–516, June 1978.

7. ——, Jr., "A logic system for fault test generation", *IEEE Trans. Comput.* Vol. C–25, No. 6, pp. 620–630, June 1976.

8. Anderson, D. A. and Metze, G., "Design of totally self-checking check circuits for *m*-out-of-*n*-codes", *IEEE Trans. Comput.*, Vol. C–22, No. 3, pp. 263–269, March 1973.

9. Armstrong, D. B., "A general method of applying error correction to synchronous digital systems", *Bell Syst. Tech. J.*, Vol. 40, pp. 577–593, March 1961.

10. ——, "On finding a nearly minimal set of fault detection tests for combinational logic nets", *IEEE Trans. Electron. Comput.*, Vol. EC–15, No. 1, pp. 66–73, Feb. 1966.

11. ——, "A deductive method for simulating faults in logic circuits", *IEEE Trans. Comput.*, Vol. C–21, No. 3, pp. 305–309, March 1972.

12. Aufenkamp, D. D. and Hohn, F. E., "Analysis of sequential machines", *IRE Trans. Electron. Comput.*, Vol. EC–6, pp. 276–285, Dec. 1957.

13. Avizienis, A., Gilley, G. C., Mathur, F. P., Rennels, D. A., Rohr, J. A. and Rubin, D. K., "The STAR (Self-testing and repairing) computer: An investigation of the theory and practice of fault-tolerant computer design", *IEEE Trans. Comput.*, Vol. C–20, No. 11, pp. 1312–1321, Nov. 1971.

14. Barraclough, W., Chiang, A. C. L. and Sohl, W., "Techniques for testing the microcomputer family", *Proc. IEEE*, Vol. 64, No. 6, pp. 943–950, June 1976.

15. Bassett, J. C. and Kime, C. R., "Improved procedure for determining diagnostic resolution", *IEEE Trans. Comput.*, Vol. C–21, No. 4, pp. 385–388, April 1972.

16. Basu, S. K., Paul, J. C. and Bhattacharjee, P. R., "Complete test-set generation for bridging faults in combinational-logic circuits", *Information Sciences*, Vol. 38, pp. 257–269, June 1986.

17. Batni, R. P. and Kime, C. R., "A module level testing approach for combinational networks", *IEEE Trans. Comput.*, Vol. C–25, No. 6, pp. 594–604, June 1976.

18. Bearnson, L. W., and Carroll, C. C., "On the design of minimum length fault tests for combinational circuits", *IEEE Trans. Comput.*, Vol. C–20, No. 11, pp. 1353–1356, Nov. 1971.

19. Bennetts, R. G., Brittle D. C., Prior, A. C. and Washington, J. L., "A modular approach to test sequence generation for large digital networks", *Digital Processes*, Vol. 1, No. 1, pp. 3–24, 1975.

20. Bennetts, R. G. and Scott, R. V., "Recent development in the theory and practice of testable logic design", *IEEE Computer*, Vol. 9, No. 6, pp. 47–63, June 1976.

21. Berger, I. and Kohavi, Z., "Fault detection in fan-out free combinational networks", *IEEE Trans. Comput.*, Vol. C–22, No. 10, pp. 908–914, Oct. 1973.

22. Bhattacharyya, A., "Very easily testable/diagnosable sequential machine", *Electronics Letters*, Vol. 15, pp. 190–191, March 1979.

23. ——, "Fault detection in Moore-model sequential machine using counter-cycle technique", *Electronics Letters*, Vol. 15, pp. 402–403, June 1979.

24. ——, "On the design of efficient checking experiments for synchronous sequential machines with failure detection capabilities", *Ph.D. dissertation*, Univ. of Calcutta for 1981 Convocation.

25. ——, "On a novel approach of fault detection in an easily testable sequential machine with extra inputs and extra outputs", *IEEE Trans. Comput.*, Vol. C–32, No. 3, pp. 323–325, March 1983.

26. Bhattacharyya, B. B., Gupta, B., Sarkar, S. and Choudhury, A. K., "Testable design of RMC networks with universal tests for detecting stuck-at and bridging faults", *IEE Proc.*, Vol. 132, Pt. E, No. 3, pp. 155–161, May 1985.

27. Bhattacharyya, D. and Hayes, J. P., "High-level test generation using bus faults", *Proc. 15th Fault-Tolerant Computing symp.*, Ann Arbor, Michigan, pp. 1–6, June 1985.

28. Bosson, D. C. and Hong, S. J., "Cause-effect analysis for multiple fault detection in combinational networks", *IEEE Trans. Comput.*, Vol. C–20, No. 11, pp. 1252–1257, Nov. 1971.

29. Booth, T. L., *Sequential machines and automata theory*, New York: John Wiley, New York, 1967.

30. Bouricius, W. G., Hsieh, E. P., Putzolu, G. R., Roth, J. P., Schneider, P. R. and Tan, C. J., "Algorithm for detection of faults in logic circuits", *IEEE Trans. Comput.*, Vol. C–20, No. 11, pp. 1258–1264, Nov. 1971.

31. Bouricius, W. G., Carter, W, C., Jessep, D. C., Schneider, P. R. and Wadia, A. B., "Reliability modeling for fault-tolerant computers", *IEEE Trans. Comput.*, Vol. C–20, No. 11, pp. 1306–1311, Nov. 1971.

32. Boute, R. T. "Distinguishing sets for optimal state identification in checking experiments", *IEEE Trans. Comput.*, Vol. C–23, No. 8, pp. 874–877, Aug. 1974.

33. ——, "Optimal and near-optimal checking experiments for output faults in sequential machines", *IEEE Trans. Comput.*, Vol. C–23, No. 11, pp. 1207–1213, Nov. 1974.

34. Braun, R. D. and Givone, D. D., "An improved algorithm for deriving checking experiments", *IEEE Trans. Comput.*, Vol. C–28, No. 2. pp. 153–156, Feb. 1979.

35. ——, "A generalization algorithm for constructing checking sequences", *IEEE Trans. Comput.*, Vol. C–30, No. 2, pp. 141–143, Feb. 1981.

36. Breuer, M. A. and Friedman, A. D., *Diagnosis and Reliable Design of Digital Systems*, Computer Science Press, 1976.

37. Breuer, M. A., "A random and algorithmic technique for fault detection test generation for sequential circuits", *IEEE Trans. Comput.*, Vol. C–20, No. 11, pp. 1364–1370, Nov. 1971.

38. ——, "Generation of fault tests for linear logic networks", *IEEE Trans. Comput.* Vol. C–21, No. 1, pp. 79–83, Jan. 1972.

39. ——, "Testing for intermittent faults in digital circuits", *IEEE Trans. Comput.*, Vol. C–22, No. 3, pp. 241–246, March 1973.

40. Carter, W. C., "Fault-tolerant Computing: An introduction and a view-point", *IEEE Trans. Comput.*, Vol. C–22, No. 3, pp. 225–229, March 1973.

41. Carroll, B. D., Shah, H. G. and Jones, D. M., "An examination of algebraic test generation methods for multiple faults", *IEEE Trans. Comput.*, Vol. C–23, No. 7, pp. 743–745, July 1974.

42. Chang, H. Y., Manning, E. G. and Metze, G., *Fault Diagnosis of Digital Systems*, New York: John Wiley, 1970.

43. Chang, H. Y., "An algorithm for selecting an optimum set of diagnostic tests", *IEEE Trans. Electron. Comput.*, Vol. EC–14, No. 5, pp. 706–711, Oct. 1965.

44. Chattopadhyay, D. K., Sengupta, A. and Choudhury, A. K., "On shift register realization of sequential machines", *Int. J. Systems Sci.*, Vol. 5, No. 11, pp. 1007–1024, 1974.

45. Chiang, A. C. L. and McCaskill, R., "Two new approaches simplify testing of microprocessors", *Electronics*, Vol. 49, No. 2, pp. 100–105, Jan. 1976.

46. Clary, J. B. and Sacane, R. A., "Self-testing computers", *IEEE Computer*, Vol. 12, No. 10, pp. 49–59, Oct. 1979.

47. Clegg, F. W., "Use of SPOOF's in the analysis of faulty logic networks", *IEEE Trans. Comput* , Vol. C–22, No. 3, pp. 229–234, March 1973.

48. Colbourne, E. D., Coverley, G. P. and Behera, S. K., "Reliability of MOS LSI circuits", *Proc. IEEE*, Vol. 62, No. 2, pp. 244–259, Feb. 1974.

49. Coy, W., "On the design of easily testable iterative systems of combinational cells", *IEEE Trans. Comput.*, Vol. C–28, No. 5, pp. 367–371, May 1979.

50. Crafts, J. M., "Techniques for memory testing", *IEEE Computer*, Vol. 12, No. 10, pp. 23–31, Oct. 1979.

51. Curtis, H. A., "A new type of double-rank sequential machine", *IEEE Trans. Comput.*, Vol. C–22, No. 3, pp. 229–234, March 1973.

52. Daehn, W. and Mucha, J., "A hardware approach of self-testing of large programmable logic arrays", *IEEE Trans. Comput.*, Vol. C–30, No. 11, pp. 829–833, Nov. 1981.

53. Dandapani, R. and Reddy, S. M., "On the design of logic networks with redundancy and testability considerations", *IEEE Trans. Comput.*, Vol. C–23, No. 11, pp. 1139–1149, Nov. 1974.

54. Das, P. and Farmer, D. E., "Fault-detection experiments for parallel decomposable sequential machines", *IEEE Trans. Comput.*, Vol. C–24, No. 11, pp. 1104–1109, Nov. 1975.

55. Das, S. R. and Sheng, C. L., "Some further studies on machine identification by merging of states", *Trans. Engineering Inst. of Canada*, Vol. 13, No. C–6, pp. I–VII, July 1970.

56. Das, S. R., "On a new approach for finding all the modified cut-sets in an incompatibility graph", *IEEE Trans. Comput.*, Vol. C–22, No. 2, pp. 187–193, Feb. 1973.

57. Das, S. R., Srimani, P. K. and Datta, C. R., "On multiple fault analysis in combinational circuits by means of boolean difference", *Proc. IEEE*, Vol. 64, No. 9, pp. 1447–1449, Sept. 1976.

58. Das, S. R., Datta, C. R., Srimani, P. K. and Mandal, K., "Comments on 'Derivation of minimal complete sets of test-input sequences using boolean differences'', *IEEE Trans. Comput.*, Vol. C–25, No. 10, pp. 1053–1056, Oct. 1976.

59. Das, S. R. and Bhattacharyya, A., "Fault detection in sequential machines with increased fault coverage", *Electronics Letters*, Vol. 14, No. 2, pp. 28–29, Jan. 1978.

60. Das, S. R., Chen, Z., Wu, S. M., Lee, S. Y. and Bhattacharyya, A., "On the design of improved failure detection experiments in synchronous sequential machines based on terminal measurements", *Computers and Electrical Engineering*, Vol. 6, No. 4, pp. 293–297, Fall 1979.

61. Das, S. R., Sheng, C. L., Chen, Z and Hsu, W. J., "Transition matrices in the measurement and control of synchronous sequential machines", *Information Sciences*, Vol. 18, No. 1, pp. 47–65, June 1979.

62. Das, S. R., Chen, Z. and Dai, Y. L., "Easily testable sequential machines with extra inputs and extra outputs", *Electronics Letters*, Vol. 16, pp. 119–121, Feb. 1980.

63. Das, S. R., Chao, P. and Chen, Z., "Transition matrices in multiple preset experiments and initial state identification of synchronous sequential machines", *Electronics Letters*, Vol. 15, No. 24, pp. 777–779, Nov. 1979.

64. Das, S. R. and Jone, W. B., "Digital signature analysis in circuit fault detection", *Proc. MICCASP–84*, Silver Jubilee Conf., Dept. of Electronics and Comm. Engg., Osmania Univ., Hyderabad, India, pp. A-79–A-81, 1984.

65. Das, S. R., "Non-transient fault testing in sequential logic circuits using stochastic modeling and simulation", *Proc. Platinum Jubilee Conf. on systems and signal processing*, Dept. of Elect. Engg., I.I.Sc. Bangalore, India, pp. 396–399, Dec. 1986.

66. Das, S. R., Jone, W. B. and Wong, K. L., "Probabilistic modeling and fault analysis in sequential logic using computer simulation", in *Computer Aided Design, Modeling and Simulation*, M. H. Hamza, Editor, Calgary, Alberta: Acta Press, pp. 87–93, 1986.

67. David, R., "Testing by feedback shift register", *IEEE Trans. Comput.*, Vol. C–29, No. 7, pp. 668–673, July 1980.

68. De Sarkar, S. C., Bandyopadhyay, S. and Choudhury, A. K., "Unate cascade realizations of synchronous sequential machines", *IEEE Trans. Comput.*, Vol. C–23, pp. 1008–1019, Oct. 1974.

69. Diaz, M., Geffroy, J. C. and Courvoisier, M., "On set realization of fail-safe sequential machines", *IEEE Trans. Comput.*, Vol. C–23, No. 2, pp. 133–138, Feb. 1974.

70. Diaz, M., Azema, P. and Ayache, J. M., "Unified design of self-checking and fail-safe combinational circuits and sequential machines", *IEEE Trans. Comput.*, Vol. C–28, No. 3, pp. 276–281, March 1979.

71. Du, M. W., "Multiple fault detection in combinational circuits: Algorithm and computational results", *IEEE Trans. Comput.*, Vol. C–22, No. 3, pp. 235–240, March 1973.

72. El-Ziq, Y. M. and Su, S. Y. H., "Fault diagnosis of MOS combinational networks", *IEEE Trans. Comput.*, Vol. C–31, No. 2, pp. 129–139, Feb. 1982.

73. Farmer, D. E., "Algorithm for designing fault-detection experiments in sequential machines", *IEEE Trans. Comput.*, Vol. C–22, No. 2, pp. 159–167, Feb. 1973.

74. Ferrari, D. and Grasselli, A., "A cellular structure for sequential networks", *IEEE Trans. Comput.*, Vol. C–18, pp. 947–953, Oct. 1969.

75. Fraley, D. and Talavage, J., "A realization algorithm for multiple preset experiments", *IEEE Trans. Comput.*, Vol. C–25, No. 11, pp. 1130–1132, Nov. 1976.

76. Freeman, H. A. and Metze, G., "Fault-tolerant Computers using 'Dotted logic' redundancy techniques", *IEEE Trans. Comput.*, Vol. C–21, No. 8, pp. 861–871, Aug. 1972.

77. Friedman, A. D., "Fault detection in redundant circuits", *IEEE Trans. Electron. Comput.* (short notes), Vol. EC–16, pp. 99–100, Feb. 1967.

78. Friedman, A. D. and Menon, P. R., *Fault Detection in Digital Circuits*, Englewood Cliffs, N.J.: Prentice Hall, 1971.

79. ——, "Restricted checking sequences for sequential machines", *IEEE Trans. Comput.*, Vol. C–22, No. 4, pp. 397–399, April 1973.

80. Friedman, A. D., "Easily testable iterative systems", *IEEE Trans. Comput.*, Vol. C–22, No. 12, pp. 1061–1064, Dec. 1973.

81. ——, "Diagnosis of short circuit faults in combinational circuits", *IEEE Trans. Comput.*, Vol. C–23, No. 7, pp. 746–752, July 1974.

82. Friedman, A. D. and Simoncini, L., "System-level fault diagnosis", *IEEE Computer*, Vol. 13, No. 3, pp. 47–53, March 1980.

83. Frohwork, "Signature analysis: A new digital field service method", *Hewlett-Packard, J.*, pp. 2–8, May 1977.

84. Fujiwara, H. and Kinoshita, K., "Design of diagnosable sequential machines utilizing extra outputs", *IEEE Trans. Comput.*, Vol. C–23. No. 2, pp. 138–145, Feb. 1974.

85. Fujiwara, H. Nagao, Y., Sasao, T. and Kinoshita, K., "Easily testable sequential machines with extra inputs", *IEEE Trans. Comput.*, Vol. C–24, No. 8, pp. 821–826, Aug. 1975.

86. Fujiwara, H. and Kinoshita, K., "On the complexity of system diagnosis", *IEEE Trans. Comput.*. Vol. C–27, pp. 881–885, Oct. 1978.

87. ——, "A design of programmable logic arrays with universal tests", *IEEE Trans. Comput.*, Vol. C–30, No. 11, pp. 823–828, Nov. 1981.

88. Fujiwara, H. and Toida, S., "The complexity of fault detection problems in combinational logic circuits", *IEEE Trans. Comput.*, Vol. C–31, No. 6, pp. 555–560, June 1982.

89. Fujiwara, H., "A new PLA design for universal testability" *IEEE Trans. Comput.*, Vol. C–33, No. 8, pp. 745–750, Aug. 1984.

90. Gault, J. W., Robinson, J. P. and Reddy, S. M., "Multiple fault-detection in combinational networks", *IEEE Trans. Comput.*, Vol. C–21, No. 1, pp. 31–36, Jan. 1972.

91. Gill, A., "State identification experiments in finite automata", *Inform. Control*, Vol. 4, pp. 132–154, 1961.

92. ——, *Introduction to the Theory of Finite State Machines*, New York: McGraw-Hill, 1962.

93. Ginsburg, S., "On the length of the smallest uniform experiment which distinguishes the terminal states of a machine", *J. ACM.*, Vol. 5, pp. 266–280, July 1958.

94. Griswold, V. M. and Carroll, C. C., "Large-scale circuit interconnection for boolean function implementation", *IEEE Trans. Comput.*, Vol. C–20, No. 5, pp. 572–575, May 1971.

95. Goldstein, L. H., "Controllability/observability analysis of digital circuits", *IEEE Trans. Circuits Syst.*, Vol. CAS–26, pp. 685–693, Sept. 1979.

96. Gönenc, "A method for design of fault detection experiments", *IEEE Trans. Comput.* (Short notes), Vol. C–19, No. 6, pp. 551–558, June 1970.

97. Goundan, A. and Hayes, J. P., "Design of totally fault locatable combinational networks", *IEEE Trans. Comput.*, Vol. C–29, No. 1, pp. 33–34, Jan. 1980.

98. Hadlock, F., "On finding a minimum set of diagnostic tests", *IEEE Trans. Electron. Comput.*, Vol. EC–16, No. 5, pp. 674–675, Oct. 1967.

99. Hartmanis, J. and Stearns, R. E., *Algebraic Structure Theory of Sequential Machines*, Englewood Cliffs, N.J.: Prentice-Hall, 1966.

100. Hayes, J. P., "A nand model for fault diagnosis in combinational logic networks", *IEEE Trans. Comput.*, Vol. C–20, No. 12, pp. 1496–1506, Dec. 1971.

101. ——, "On realizations of boolean functions requiring a minimal and near-minimal number of tests", *IEEE Trans. Comput.*, Vol. C–20, No. 12, pp. 1506–1513, Dec. 1971.

102. ——, "On modifying logic networks to improve their diagnosability", *IEEE Trans. Comput.*, Vol. C–23, No. 1, pp. 56–62, Jan. 1974.

103. Hayes, J. P. and Friedman, A. D., "Test point placement to simplify fault detection", *IEEE Trans. Comput.*, Vol. C–23, No. 7, pp. 727–735, July 1974.

104. Hayes, J. P., "Detection of pattern sensitive faults in random-access memories", *IEEE Trans. Comput.*, Vol. C–24, No. 2, pp. 150–157, Feb. 1975.

105. ——, "Transition count testing of combinational logic circuits", *IEEE Trans. Comput.*, Vol. C–25, No. 6, pp. 613–620, June 1976.

106. ——, "A graph model for fault-tolerant computing systems", *IEEE Trans. Comput.*, Vol. C–25, No. 9, pp. 875–884, Sept. 1976.

107. ——, "On the properties of irredundant logic networks", *IEEE Trans. Comput.*, Vol. C–25, No. 9. pp. 884–892, Sept. 1976.

108. ——, "Generation of optimal transition count test", *IEEE Trans. Comput.*, Vol. C–27, No. 1, pp. 36–41, Jan. 1978.

109. ——, *Computer Architecture and Organization*: New York: McGraw-Hill, 1978.

110. Hayes, J. P. and McCluskey, E. J., "Testability considerations in microprocessor based design", *IEEE Computer*, Vol. 13, No. 3, pp. 17–26, March 1980.

111. Hayes, J. P., "Testing memories for a single cell pattern sensitive faults", *IEEE Trans. Comput.*, Vol. C–29, No. 3, pp. 249–254, March 1980.

112. ——, "A survey of bit-sliced computer design", *J. Digital Systems*, Vol. 5, pp. 203–250, Fall 1981.

113. Hennie, F. C., "Fault detecting experiments for sequential circuits", *Proc. 5th Annual Symp. Switching Theory and Logical Design*. Princeton, N. J.: Princeton Univ. Press, pp. 95–110, 1964.

114. ——, *Finite State Models for Logical Machines*, New York: John Wiley, 1968.

115. Hibbard, T. N., "Least upper bounds on minimal terminal state experiments for two classes of sequential machines", *J. ACM.*, Vol. 8, pp. 601–612, Oct. 1961.

116. Holborow, C. E., "An improved bound on the length of checking experiments for sequential machines with counter-cycles", *IEEE Trans. Comput.*, Vol. C–21, No. 6, pp. 597–598, June 1972.

117. Hong, S. J., "Existence algorithm for synchronizing/distinguishing sequences", *IEEE Trans. Comput.*, Vol. C–30, No. 3, pp. 234–237, March 1981.

118. Hohn, F. E., Seshu, S. and Aufenkamp, D. D., "The theory of nets", *IRE Trans. Electron. Comput.*, Vol. EC–6, pp. 154–161, Sept. 1957.

119. Hornbuckle, G. D. and Span, R. N.. "Diagnosis of single-gate failures in combinational circuits", *IEEE Trans. Comput.*, Vol. C–18, No. 3, pp. 216–220, March 1969.

120. Hsieh, E. P., "Checking experiments for sequential machines", *IEEE Trans. Comput.*, Vol. C–20, No. 10, pp. 159–161, Oct. 1971.

121. Joseph, J., "On easily diagnosable sequential machines", *IEEE Trans. Comput.*, Vol. C–27, No. 2, pp. 159–161, Feb. 1978.

122. Khakbaz, J., "Totally self-checking checker for 1-out-of-*n* code using two rail codes", *IEEE Trans. Comput.*, Vol. C–21, No. 7, pp. 677–681, July 1982.

123. ——, "A testable PLA design with low overhead and high fault coverage", *IEEE Trans. Comput.*, Vol. C–33, No. 8, pp. 743–745, Aug. 1984.

124. Kamal, S. and Page, C. V., "Intermittent faults: A model and a detection procedure", *IEEE Trans. Comput.*, Vol. C–23, No. 7, pp. 713–719, July 1974.

125. Kane, J. R. and Yau, S. S., "On the design of easily testable machines", in *IEEE 12th Annual Symp. Proc. Switching and Automata Theory*, pp. 38–42, Oct. 1971.

126. Kautz, W. H., "Fault testing and diagnosis in combinational digital circuits", *IEEE Trans. Comput.*, Vol. C–17, No. 4, pp. 352–366, April 1968.

127. Kella, J., "Sequential machine identification", *IEEE Trans. Comput.*, Vol. C–20, No. 3, pp. 332–338, March 1971.

128. Kime, C. R., "An organization for checking experiments on sequential circuits", *IEEE Trans. Electron. Comput.* (Short notes), Vol. EC–15, No. 1, pp. 113–115, Feb. 1966.

129. ——, "An analysis model for digital system diagnosis", *IEEE Trans. Comput.*, Vol. C–19, No. 11, pp. 1063–1969, Nov. 1970.

130. ———, "Fault-tolerant computing: An introduction and a perspective", *IEEE Trans. Comput.*, Vol. C–24, No. 5, pp. 457–460, May 1975.

131. Kodandapani, K. L., "A note on easily testable realizations for logic functions", *IEEE Trans. Comput.*, Vol. C–23, No. 3, pp. 332–333, March 1974.

132. Kodandapani, K. L. and Pradhan, D. K., "Undetectability of bridging faults and validity of stuck at fault test sets", *IEEE Trans. Comput.*, Vol. C–29, No. 1, pp. 55–59, Jan. 1980.

133. Kohavi, Z. and Lavalee, P., "Design of sequential machines with fault detection capabilities", *IEEE Trans. Electron. Comput.*, Vol. EC–16, No. 4, pp. 473–484, Aug. 1967.

134. Kohavi, I. and Kohavi, Z., "Variable length distinguishing sequences and their application to the design of fault-detection experiments", *IEEE Trans. Comput.*, (Short notes), Vol. C–17, No. 8, pp. 792–795, Aug. 1968.

135. Kohavi, Z., *Switching and Finite Automata Theory*, New York: McGraw-Hill, 1970.

136. Kohavi, Z. and Spires, D. A., "Designing sets of fault detection tests for combinational logic circuits", *IEEE Trans. Comput.*, Vol. C–20, No. 12, pp. 1463–1469, Dec. 1971.

137. Kohavi, I. and Kohavi, Z., "Detection of multiple faults in combinational logic networks", *IEEE Trans. Comput.*, Vol. C–21, No. 6, pp. 556–568, June 1972.

138. Kohavi, Z. and Berger, I, "Fault diagnosis in combinational tree networks", *IEEE Trans. Comput.*, Vol. C–24, No. 12, pp. 1161–1167, Dec. 1975.

139. Koren, I. and Kohavi, Z., "Diagnosis of intermittent faults in combinational networks", *IEEE Trans. Comput.*, Vol. C–26, No. 11, pp. 1154–1158, Nov. 1977.

140. Ku, C. T. and Masson, G. M., "The boolean difference and multiple fault analysis", *IEEE Trans. Comput.*, Vol. C–24, No. 1, pp. 62–71, Jan. 1975.

141. Lala. P. K., *Fault Tolerant and Fault Testable Hardware Design*: London: Prentice-Hall International, 1985.

142. Larson, R. W. and Reed, I. S., "Redundacy by coding versus redundancy by replication for failure-tolerant sequential circuits", *IEEE Trans. Comput.*, Vol. C–21, No. 2, pp. 130–137, Feb. 1972.

143. Lee, S. C., *Digital Circuits and Logical Design*, Englewood Cliffs, N.J.: Prentice-Hall, 1976.

144. Liaw, C. C. and Su, S. Y. H., "Test-experiments for detection and location of intermittent faults in sequential circuits", *IEEE Trans. Comput.*, Vol, C–30, No. 12, pp. 989–995, Dec. 1981.

145. Lyons, R. E. and Vanderkulk, W., "The use of triple-modular redundancy to improve computer reliability", *IBM J. Res. and Develop.*, Vol. 6, pp. 200–209, April 1962.

146. Mandelbaum, D., "On error control in sequential machines" *IEEE Trans. Comput.*, Vol. C–21, No. 5, pp. 492–495, May 1972.

147. Maki, G. K. and Sawin, III, D. H., "Fault-tolerant asynchronous sequential machines", *IEEE Trans. Comput.*, Vol. C–23, No. 7, pp. 651–657, July 1974.

148. Marinos, P. N., "Derivation of minimal complete sets of test-input sequences using boolean differences", *IEEE Trans. Comput.*, Vol. C–20, No. 1, pp. 25–32, Jan. 1971.

149. Mathur, F. P., "On reliability modelling and analysis of ultrareliable fault-tolerant digital system", *IEEE Trans. Comput.*, Vol. C–20, No. 11, pp. 1376–1382, Nov. 1971.

150. Mayer, J. F., "Fault tolerant sequential machine", *IEEE Trans. Comput.* Vol. C–20, No. 10, pp. 1167–1177, Oct. 1971.

151. Mayer, J. F. and Sundstorm, R. J., "On-line diagnosis of unrestricted faults", *IEEE Trans. Comput.*, Vol. C–24, No. 5, pp. 468–475, May 1975.

152. McCluskey, E. J., *Introduction to the Theory of Switching Circuits*, New York: McGraw-Hill, 1965.

153. McCluskey, E. J. and Clegg, F. W., "Fault equivalence in combinational logic networks", *IEEE Trans. Comput.*, Vol. C–20, No. 11, pp. 1286–1293, Nov. 1971.

154. ———, "Fault equivalence in combinational logic networks", *IEEE Trans. Comput.*, Vol. C–23, No. 7, pp. 720–727, July 1974.

155. McPherson, J. A. and Kime, C. R., "A two level diagnostic model for digital systems", *IEEE Trans. Comput.*, Vol. C–28, No. 1, pp. 16–27, Jan. 1979.

156. Mine, H. and Koga, Y., "Basic properties and a construction method for fail-safe logic system", *IEEE Trans. Electron. Comput.*, Vol. EC–16, pp. 282–289, June 1967.

157. Moore, E. F., "Gedanken-experiments on sequential machiness" in *Automata Studies* (Annals of Mathematical Studies No. 34), C. E. Shanon and J. McCarthy, Edition, Princeton, N.J.: Princeton Univ. Press, pp. 129–153, 1956.

158. Muehldorf, E. I. and Savkar, A. D., "LSI logic testing—an overview", *IEEE Trans. Comput.*, Vol. C–30, No. 1, pp. 1–16, Jan. 1981.

159. Mukhopadhyay, A. and Schmitz, G., "Minimization of exclusive-OR and logical equivalence switching circuits", *IEEE Trans. Comput.*, Vol. C–19, No. 2, pp. 132–140, Feb. 1970.

160. Murakami, S. I., Kinoshita, K. and Ozaki, Z., "Sequential machines capable of fault diagnosis", *IEEE Trans. Comput.*, Vol. C–19, No. 11, pp. 1079–1085, Nov. 1970.

161. Osman, M. Y. and Dennis Weiss, C., "Shared logic realizations of dynamically self-checked and fault-tolerant logic", *IEEE Trans. Comput.*, Vol. C–22, No. 3, pp. 298–306, March 1973.

162. Ostapko, D. L. and Hong, S. J., "Fault analysis and test generation for programmable logic arrays (PLA's)", *IEEE Trans. Comput.*, Vol. C–28, No. 9, pp. 617–627, Sept. 1979.

163. Page, E. W. and Marinos, P. N., "Programmable array realizations of synchronous sequential machines", *IEEE Trans. Comput.*, Vol. C–26, No. 8, pp. 811–818, Aug. 1977.

164. Paige, M. R., "Synthesis of diagnosable FET networks", *IEEE Trans. Comput.*, Vol. C–22, No. 5, pp. 513–515, May 1973.

165. Papachristou, C. A. and Sarma, D., "An approach to sequential circuit construction in LSI programmable arrays", *IEE Proc.*, Vol. 130, Pt. E, No. 5, pp. 159–164, Sept. 1983.

166. Parthasarathy, R. and Reddy, S. M., "A testable design of iterative logic arrays", *IEEE Trans. Comput.*, Vol. C–30, No. 11, pp. 833–841, Nov. 1981.

167. Pierce, W. H., *Failure-tolerant Computer Design*, New York: Academic Press, 1965.

168. Poage, J. F., "Derivation of optimal test to detect faults in combinational circuits", *Proc. Symp. Mathematical Theory of Automation*, Polytechnic Inst. of Brooklyn, pp. 483–528, 1963.

169. Poage, J. F. and McCluskey, E. J., "Derivation of optimum test sequences for sequential machines", *Proc. 5th Annual Symp. Switching Circuit Theory and Logical Design*, Princeton, N.J.: Princeton Univ. Press, pp. 121–132, 1964.

170. Powell, T. J., "A procedure for selecting diagnostic tests", *IEEE Trans. Comput.*, Vol. C–22, No. 7, pp. 662–668, July 1973.

171. Pradhan, D. K. and Reddy, S. M., "Fault-tolerant asynchronous networks", *IEEE Trans. Comput.*, Vol. C–22, No. 7, pp. 662–668, July 1973.

172. Pradhan, D. K. and Stifter, J. J., "Error correcting codes and self-checking circuits", *IEEE Computer*, Vol. 13, No. 3, pp. 27–38, March 1980.

173. Pradhan, D. K., "Sequential network design using extra inputs for fault detection", *IEEE Trans. Comput.*, Vol. C–32, No. 3, 319–323, March 1983.

174. ———, "Fault-tolerant multiprocessor link and bus network architectures", *IEEE Trans. Comput.*, Vol. C–34, No. 1, pp. 33–45, Jan. 1985.

175. ———, *Fault-Tolerant Computing*, Vol. I and Vol. II, Englewood Cliffs, N.J.: Prentice-Hall, 1986.

176. Ramamoorthy, C. V. and Chang, L. C., "System segmentation for parallel diagnosis of computers", *IEEE Trans. Comput.*, Vol. C–20, No. 3, pp. 261–270, March 1971.

177. Ramamoorthy, C. V., "Fault-tolerant computing: An introduction and an overview", *IEEE Trans. Comput.*, Vol. C–20, No. 11, pp. 1241–1243, Nov. 1971.

178. Ramamoorthy, C. V. and Mayeda, W., "Computer diagnosis using blocking gate approach", *IEEE Trans. Comput.*, Vol. C–20, No. 11, pp. 1294–1299, Nov. 1971.

179. Ramanatha, K. S., and Biswas, N. N., "A design for testability of undetectable crosspoint faults in programmable logic arrays", *IEEE Trans. Comput.*, Vol. C–32, No. 6, pp. 551–557, June 1983.

180. Reddy, S. M., "A design procedure for fault locatable switching circuits", *IEEE Trans. Comput.*, Vol. C–21, No. 12, pp. 1421–1426, Dec. 1972.

181. ———, "Complete test sets for logic functions", *IEEE Trans. Comput.* Vol. C–22, No. 11, pp. 1016–1020, Nov. 1973.

182. ———, "A note on self-checking checkers", *IEEE Trans. Comput.*, Vol. C–23, No. 10, pp. 1100–1102, Oct. 1974.

183. ———, "Easily testable realizations for logic functions", *IEEE Trans. Comput.*, Vol. C–21, No. 11, pp. 1183–1188, Nov. 1974.

184. Roth, J. P., "Diagnosis of automata failures: A calculus and a method", *IBM J. Res. Develop.*, Vol. 10, pp. 278–294, July 1966.

185. Roth, J. P., Bouricius, W. G. and Schneider, P. R., "Programmed algorithms to compute tests to detect and distinguish between failures in logic circuits", *IEEE Trans. Electron Comput.*, Vol. EC–16, No. 5, pp. 567–580, Oct. 1967.

186. Russell, J. D. and Kime, C. R., "Structural factors in the fault diagnosis of combinational networks", *IEEE Trans. Comput.*, Vol. C–20, No. 11, pp. 1276–1285, Nov. 1971.

187. ———, "System fault diagnosis: Closure and diagnosibility with repair", *IEEE Trans. Comput.*, Vol. C–24, No. 11, pp. 1079–1088, Nov. 1975.

188. ———, "System fault diagnosis: Masking, exposure, and diagnosibility with repair", *IEEE Trans. Comput.*, Vol. C–24, No. 12, pp. 1155–1161, Dec. 1975.

189. Saluja, K. K., "Synchronous sequential machines: A modular and testable design", *IEEE Trans. Comput.*, Vol. C–29, No. 11, pp. 1020–1025, Nov. 1980.

190. Saluja, K. K., Kinoshita, K. and Fujiwara, H., "An easily testable design of programmable logic arrays for multiple faults", *IEEE Trans. Comput.*, Vol. C–32, No. 11, pp. 1038–1046, Nov. 1983.

191. Saluja, K. K. and Kinoshita, K., "Test pattern generation for API faults in RAM", *IEEE Trans. Comput.*, Vol. C–34, No. 3, pp. 284–287, March 1985.

192. Saluja, K. K. and Dandapani, R., "An alternative to scan design methods for sequential machines", *IEEE Trans. Comput.*, Vol. C–35, No. 4, pp. 384–388, April 1986.

193. Sawin III, D. H., "Design of reliable synchronous sequential circuits", *IEEE Trans. Comput.*, Vol. C–24, No. 5, pp. 567–570, May 1975.

194. Schertz, D. R. and Metze, G., "A new representation for faults in combinational digital circuits", *IEEE Trans. Comput.*, Vol. C–21, No. 8, pp. 858–866, Aug. 1972.

195. Schneider, P. R., "The necessity to examine *D*-chains in diagnostic test generation—an example", *IBM, J. Res. Develop.*, Vol. 11, p. 114, Jan. 1967.

196. Sellers, E. F., Hsiao, M. Y. and Bearnson, L. W., "Analysing errors with the boolean difference", *IEEE Trans. Comput.*, Vol. C–17, No. 7, pp. 676–683, July 1968.

197. Sengupta, A., Chattopadhyay, D. K., Palit, A., Bandyopadhyay, A. K., Basu, M. S. and Choudhury, A. K., "Realization of fault-tolerant and fail-safe sequential machines", *IEEE Trans. Comput.*, Vol. C–26, No. 1, pp. 91–96, Jan. 1977.

198. Sengupta, A., Chattopadhyay, D. K., Palit, A., Bandyopadhyay, A. K. and Choudhury, A. K., "Realization of fault-tolerant machines—Linear code application", *IEEE Trans. Comput.*, Vol. C–30, No. 3, pp. 237–240, March 1981.

199. Servit, M., "Hazard correction in synchronous sequential circuits", *IEEE Trans. Comput.*, Vol. C–24, No. 3, pp. 305–310, March 1975.

200. Seshu, S., Miller, R. E. and Meize, G., "Transition matrices of sequential machines", *IRE Trans. Circuit Theory*, Vol. CT–6, pp. 5–12, March 1959.

201. Seshu, S. and Freeman, D. N., "The diagnosis of asynchronous sequential switching systems", *IRE Trans. Electron. Comput.*, Vol. EC–11, No. 4, pp. 459–465, Aug. 1962.

202. Seth, S. C. and Narayanaswamy, "A graph model for pattern sensitive faults in random access memories", *IEEE Trans. Comput.*, Vol. C–30, No. 12, pp. 973–977, Dec. 1981.

203. Shedletsky, J. J. and McCluskey, E. J., "The error latency of a fault in a sequential digital circuit", *IEEE Trans. Comput.*, Vol. C–25, No. 6, pp. 655–659, June 1976.

204. Sheng, C. L., *Introduction to Switching Logic*, Intext Educational Publishers, 1972.

205. Sheng, C. L. and Das, S. R.: "Identification of synchronous sequential machines by merging of states", *South Western IEEE Conf. Record 69C16–SWIECO*, pp. 4F1–4F8, April 1969.

206. ———, "On identification of synchronous sequential machines", *Automatica*, col. 8, pp. 357–360, May 1972.

207. Sheppard, D. A. and Vranesic, Z. G., "Fault detection of binary sequential machines", *IEEE Trans. Comput.*, Vol. C–23, No. 4, pp. 352–358, April 1974.

208. Smith, J. E. and Metze, M., "Strongly fault secure logic networks", *IEEE Trans. Comput.*, Vol. C–27, No. 6, pp. 491–499, June 1978.

209. Smith, J. E., "Measures of the effectiveness of fault signature analysis", *IEEE Trans. Comput.*, Vol. C–29, No. 6, pp. 510–514, June 1980.

210. Sinha, B. P. and Bhattacharyya, B. B., "On numerical complexity of short-circuit faults in logic networks", *IEEE Trans. Comput.*, Vol. C–34, No. 2, pp. 186–190, Feb. 1985.

211. Sridhar, T. and Hayes, J. P., "A functional approach to testing bit-sliced microprocessors", *IEEE Trans. Comput.*, Vol. C–30, No. 8, pp. 563–571, Aug. 1981.

212. ———, "Design of easily testable bit-sliced systems", *IEEE Trans. Comput.*, Vol. C–30, No. 11, pp. 842–854, Nov. 1981.

213. Su, S. Y. H. and Cho, Y.-C., "A new approach to the fault location of combinational circuits", *IEEE Trans. Comput.*, Vol. C–21, No. 1, pp. 21–30, Jan. 1972.

214. Suk, D. S. and Reddy, S. M., "A march test for functional faults in semiconductor random access memories", *IEEE Trans. Comput.*, Vol. C–30, No. 12, pp. 982–985, Dec. 1981.

215. Takaoka, T. and Mine, H., "*N*-Fail-safe logical system", *IEEE Trans. Comput.* Vol. C–20, No. 5, pp. 536–542, May 1971.

216. Tohma, Y. and Aoyage, S., "Failure-tolerant sequential machines with past information", *IEEE Trans. Comput.*, Vol. C–20, No. 4, pp. 392–396, April 1971.

217. Tohma, Y., Ohyama, Y. and Sakai, R., "Realization of fail-safe sequential machines by using a *k*-out-of-*n*-code", *IEEE Trans. Comput.*, Vol. C–20, No. 11, pp. 1270–1275, Nov. 1971.

218. Tohma, Y., "Design technique of fail-safe sequential circuits using flip slops for internal memory", *IEEE Trans. Comput.*, Vol. C–23, No. 11, pp. 1149–1154 Nov. 1974.

219. Tokura, N., Kasami, T. and Hashimoto, A., "Fail-safe logic nets", *IEEE Trans. Comput.*, Vol. C–20, No. 3, pp. 323–330, March 1971.

220. Veelenturf, L. P. J., "Inference of sequential machines from sample computations", *IEEE Trans. Comput.*, Vol. C–27, No. 2, pp. 167–170, Feb. 1978.

221. Wang, D. T., "An algorithm for the generation of test sets for combinational logic networks", *IEEE Trans. Comput.*, Vol. C–24, No. 7, pp. 742–746, July 1975.

222. ———, "Properties of faults and criticalities of values under test for combinational networks", *IEEE Trans. Comput.*, Vol. C–24, No. 7, pp. 746–750, July 1975.

223. Wakerly, J. F., "Partially self-checking circuits and their use in performing logical operations", *IEEE Trans. Comput.*, Vol. C–23, No. 7, pp. 658–666, July 1974.

224. ———, "Microcomputer reliability improvement using triple-modular redundancy", *IEEE Proc.*, Vol. 64, pp. 889–895, June 1976.

225. Wakimura, Y. and Yoshida, N., "Comments on 'checking experiments for sequential machines' ", *IEEE Trans. Comput.*, Vol. C–24, No. 11, pp. 1141–1142, Nov. 1975.

226. Williams, M. J. Y. and Angell, J. B., "Enhancing testability of large-scale integrated circuits via test points and additional logic", *IEEE Trans. Comput.*, Vol. C–24, No. 1, pp. 46-60, Jan. 1973.

227. Williams, T. W. and Parker, K. P., "Testing logic networks and designing for testability", *IEEE Computer*, Vol. 12, No. 10, pp. 9–21, Oct. 1979.

228. ——, "Design for testability—A survey", *IEEE Trans. Comput.*, Vol. C–31, No. 1., pp. 2–15, Jan. 1982.

229. Yau, S. S. and Tang, Y. S., "An efficient algorithm for generating complete test sets for combinational logic circuits", *IEEE Trans. Comput.*, Vol. C–20, No. 11, pp. 1245–1251, Nov. 1971.

230. Yau, S. S. and Yang, S.-C., "Multiple fault detection for combinational logic circuits", *IEEE Trans. Comput*. Vol. C–24, No. 3, pp. 233–242, March 1975.

# Glossary

## 1. Sequential machine

A *sequential machine M* is a quintuple $(I, S, O, f, g)$, where

$I$ is a finite *input alphabet*,

$S$ is a finite *state alphabet*,

$O$ is a finite *output alphabet*,

$f$ is a *next-state function*,

$g$ is an *output function*.

All these require more clarification as follows:

The set of values which a set of variables can assume is called an *alphabet*, and each element of the alphabet is called a *symbol*. Let there be $n$ input variables $\{x_i\}$, $i = 1, \ldots, n$; $m$ state variables $\{y_i\}$, $i = 1, \ldots, m$; $l$ output variables $\{Z_i\}$, $i = 1, \ldots, l$. Thus an $n$-tuple of $n$ input variables is called an *input symbol*, and there are $2^n$ input symbols in the *input alphabet*. Similarly, an *m-tuple* of the $m$ state variables and an $l$-tuple of $l$ output variables are called *state symbol* and *output symbol* respectively. Also, there are $2^m$ state symbols in the *state alphabet* and $2^l$ output *symbols* in the *output alphabet* respectively. For simplicity, an input symbol is called an *input* and denoted by $I_i$, $i = 1, \ldots, 2^n = p$; and similarly, a state symbol is called a *state* and denoted by $S_i$, $i = 1, \ldots, 2^m = q$; and likewise, an output symbol is represented as an output and denoted by $O_i$, $i = 1, \ldots, 2^l = r$. Therefore we have,

$$I = \{I_1, \ldots, I_p\}, \text{ the input alphabet}$$

$$S = \{S_1, \ldots, S_q\}, \text{ the state alphabet}$$

$$O = \{O_1, \ldots, O_r\}, \text{ the output alphabet.}$$

Also we have $f$ the next-state mapping, is denoted as $f : S \times I \rightarrow S$ and is defined as

$$S(t_{k+1}) = f\{S(t_k), I(t_k)\}; \tag{A}$$

and $g$, the output mapping, is denoted as $g : S \times I \rightarrow O$ and is defined as

$$O(t_k) = g\{S(t_k), I(t_k)\}; \tag{B}$$

Here $t_k$ and $t_{k+1}$ denote the present and next time instant respectively. In sequential machine the values of the variables are usually specified at certain *discrete* time instants rather than over the whole continuous time. The functions $f$ and $g$ are sometimes called the *characterizing functions* and equations (A) and (B) are called *characterizing equations* of a sequential machine.

**2. Sequential circuit**

At present in the literature, the terms *sequential circuit* and *sequential machine* are used synonymously. Originally, the term machine was used to emphasize the mathematical relationship among the inputs, outputs, and states, whereas the term circuit used to emphasize the physical realization.

**3. Synchronous and asynchronous sequential machine**

A sequential machine (operating in the *pulse mode*) is said to be *synchronous* sequential machine if the following conditions are satisfied:

(i) At least one of the inputs is a pulse signal,

(ii) changes in internal state occur only in response to the occurence of a pulse at one of the pulse inputs, and each input (pulse input) causes one change in internal state.

On the other hand, a sequential machine (operating in the *fundamental mode*) is said to be *asynchronous* sequential machine if and only if the inputs are never changed unless the circuit is *stable* internally. The circuit is said to be stable internally if none of the internal signals is unstable or changing.

**4. Flow table**

A *flow table* is simply representation of the characterizing functions $f$ and $g$ in a tabular form. Thus flow table lists the values of these two functions for all state and input combinations. The output can be associated either with the present state or with the transition from the present state to the next state.

**5. Mealy machine**

In a *Mealy model* machine (or *Mealy type*) or simply *Mealy* machine, the next state is dependent on the present state and the present input, and the present output is also dependent on the present state and the present input. Thus the output variables are functions of the input variables as well as the state variables.

**6. Moore machine**

In a *Moore model* machine (or *Moore type*) or simply *Moore* machine, the next state is dependent on the present state and the present input, but the present output is dependent only on the present state. Thus the output variables are functions of the state variables only.

### 7. The state diagram

The *state diagram* of a synchronous sequential machine is a direct translation of the flow table into diagram form. The state diagram is sometimes called *transition diagram* and flow table representation is sometimes called *transition table* or *state table* representation.

### 8. Deterministic sequential machine

A sequential machine is said to be *deterministic* if and only if the future behaviour of the machine can be completely determined by the present state and the input sequence to be applied.

### 9. Completely specified sequential machine

A sequential machine $M$ is said to be *completely specified* if the next-state and present-output entries of the flow table are completely filled up, i.e., if the machine $M$ is in any state and any input is applied to it, the $M$ will go to a specified state with a specified output.

### 10. Incompletely specified sequential machine

A flow table of the sequential machine $M$ with one or more unspecified entries is called an *incompletely specified flow table* and the sequential machine with one or more unspecified entries in the flow table is called an *incompletely specified sequential machine*.

### 11. Reduced sequential machine

A *reduced* sequential machine is represented by a minimal form of the flow table where no state is *equivalent* to any other state in the flow table. A state $S_i$ of a machine $M_1$ and the state $S_j$ of the same machine or any other machine $M_2$ are said to be equivalent, if and only if $M_1$ in state $S_i$ and $M_1$ or $M_2$ in state $S_j$, when excited by any input sequence, yield identical output sequences.

### 12. Strongly connected sequential machine

A sequential machine $M$ is said to be a *strongly connected* sequential machine if and only if it is possible to reach any state $S_j$ from any other state $S_i$ or in other words, if and only if any state is accessible from every other state.

### 13. Submachine of a sequential machine

A *submachine* $M' = (I', S', O', f, g)$ of a machine $M = (I, S, O, f, g)$ is a sequential machine with $I' \subset I$, $S' \subset S$, $O' \subset O$, and $f' = f$ restricted to $S' \times I'$ and $g' = g$ restricted to $S' \times I'$.

In most cases we only consider those submachines with $I' = I$, $O' = O$, but $S' \subset S$.

**14. Isomorphic sequential machine**

The terminal behaviour of a sequential machine is not influenced by the names given to the states of the machine in its flow table. Technically, two flow tables or sequeutial machines that can be obtained from one another by renaming the states are said to be *isomorphic.*

**15. Iterative logic arrays (ILA)**

For analysis purposes, bit-sliced system can be viewed as *iterative logic arrays*, which are one-dimensional cascades of identical cells. The properties of *ILA's* of simple cells are discussed in references [78], [114].

# Index